THE LAST ASYLUM

By the same author

Eve and the New Jerusalem: Socialism and Feminism in
the Nineteenth Century
Mary Wollstonecraft and the Feminist Imagination
On Kindness (*with Adam Phillips*)
Women, Gender and Enlightenment, 1650–1850 (*ed. with Sarah Knott*)
History & Psyche: Culture, Psychoanalysis and the Past
(*ed. with Sally Alexander*)

THE LAST ASYLUM

A Memoir of Madness in Our Times

Barbara Taylor

The University of Chicago Press

CHICAGO

BARBARA TAYLOR is professor of humanities at Queen Mary University of London. She is the author of *Eve and the New Jerusalem* and *Mary Wollstonecraft and the Feminist Imagination* and coauthor, with Adam Phillips, of *On Kindness*.

The University of Chicago Press, Chicago 60637
© 2015 by Barbara Taylor
All rights reserved. Published 2015.
Printed in the United States of America

24 23 22 21 20 19 18 17 16 15   1 2 3 4 5

ISBN-13: 978-0-226-27392-1 (paper)
ISBN-13: 978-0-226-27408-9 (e-book)
DOI: 10.7208/chicago/9780226274089.001.0001

Library of Congress Cataloging-in-Publication Data
Taylor, Barbara, 1950 April 11– author.
  The last asylum : a memoir of madness in our times / Barbara Taylor.
    pages ; cm
  Includes bibliographical references and index.
  ISBN 978-0-226-27392-1 (pbk. : alk. paper) — ISBN 978-0-226-27408-9 (e-book)
  1. Taylor, Barbara, 1950 April 11—Mental health. 2. Mental illness. 3. Mentally ill women—Institutional care—Great Britain. 4. London (England). County Lunatic Asylum, Colney Hatch. 5. Friern Hospital. 6. Psychiatric hospitals—England—London. I. Title.
  RC451.4.W6T38 2015
  362.2'109421—dc23
                                                                      2014035715

For Lise

CONTENTS

PART TWO

PART THREE

PROLOGUE

November 2002. I put on a skirt and jacket and head up to Friern Barnet, in the north London suburbs. Princess Park Manor is a palatial apartment complex on the south side of Friern high street. The sharp-suited woman in the sales hut by the Manor's front gate seems excited by my visit. Maybe sales have been slow? This is not a well-heeled neighbourhood; will anyone who can pay a half-million for a two-bedroom flat really want to live here? I flip through the glossy brochures, check the location of the show apartments, and head along a flower-lined path towards the main building. The wide lawn glistens in the morning light; a squirrel chatters in a cedar. When I am out of sight of the sales hut I leave the path and walk across the lawn to a door on the building's far right. The door opens into a sunny corridor with polished wooden floors. For a moment I'm disoriented, then I walk along the corridor until a staircase appears on my left. I climb to the first floor, on to a small landing with a numbered door. I go over to the door and stand there. Sounds from within? No, nothing. No footsteps, no cries, no rattle of keys . . . Silence. So we have really been exorcized then? Not even the echoes of our voices trapped in the walls? Surely the stench of our cigarettes, our endless cigarettes, must still be detectable? But all I can smell is fresh paint and floor polish.

I go back downstairs and make my way to a couple of show flats. At first I can't understand what I find in them: galleries going nowhere, cubbyhole bedrooms with arched ceilings, windows spanning two floors, lots of odd corners. Many of the rooms are womb-like: long, narrow entrances abruptly opening into windowless spaces. Everything is decorated in mass-market chic: dark green leather sofas; marble busts on glass shelves; poster-shop prints in shiny frames. Gradually, like reassembling a picture puzzle, I begin to make out a bit of corridor here, an original ceiling there. I start to see how it's been done – and abruptly lose interest. I go back to the sales hut, smile discouragingly at the sharp suit, and head back home.

Princess Park Manor was my loony bin, my nuthouse. In the late 1980s I spent nearly eight months there. None of the sales literature mentions the Manor's previous incarnation as Friern Hospital, yet in its own way it was top-drawer even then. Entering it for the first time in July 1988, I found myself in what had once been England's biggest and most advanced psychiatric institution: the Middlesex County Pauper Lunatic Asylum at Colney Hatch, or 'Colney Hatch' as Friern was universally known until the mid-twentieth century. Like nearly all of Britain's old asylums, the hospital was slated for closure. A few days after my admission, my friend the historian Raphael Samuel had come to visit me. He looked around the vast, drab ward with undisguised fascination and embraced me. 'Darling Barbara! What a privilege for you, as a historian, to be present at the demise of one of the last great Victorian institutions!' His words both annoyed and amused me, but they lodged inside me – and this book is the result.

I had always been unhappy. Moving from Canada to London, aged twenty-one, I had thought to leave my unhappiness behind me. But it travelled with me, and in my late twenties it took a frightening turn. I became horribly anxious, unable to sleep, plagued by mysterious bodily symptoms. In 1981, when I was thirty-one, I broke down entirely for a time and decided to seek help. Many of my friends were in psychotherapy, so this became my route. In 1982 I began seeing a psychoanalyst. Within three years I had lost all semblance of ordinary life; three years further on I was admitted to Friern. I spent nearly four years in the mental health system, either as an inpatient or a day patient. While I was in Friern I lost my home, and went to live in a psychiatric hostel. By then my world had contracted around my illness and I believed I would be a lifelong mental patient.

This was not to be my fate. But if it had been, what then? What would the rest of my life have been like?

This book is, in part, an attempt to answer this question. My years as a mental patient coincided with a revolution in the psychiatric health system. Right across the western world, people with mental disorders were being decanted out of the old asylums into the 'community'. I was formally discharged in 1992. Friern closed the following year. By the end of the decade nearly all the mental hospitals had gone. I had lived through the twilight days of the Asylum Age. I was half aware of this at the time, and occasionally felt very anxious about what would become of me. But mostly these changes, inasmuch as I was conscious of them, just fed my sense of personal despair. After all, I was doomed anyway, so this hospital, that day centre, no hospital, no day centre – what difference would any of it make to me?

Yet unlike most people who lived through this transition, I had another resource to hand. All the time I was in psychiatric institutions I was also continuing to see my psychoanalyst, five times a week. This was a fraught therapeutic relationship which in the summer of 1990 went through a crisis that proved transformative. Gradually, with much grizzling and many back-steps, I moved out of the hell that I had been inhabiting. In 1993 I got a job at a London university; in the late 1990s I acquired a partner and a family, and found myself with a life that I wanted to lead. I have been living it ever since.

The Last Asylum is the story of my madness years, set inside the story of the death of the asylum system in the late twentieth century. The book is a historical meditation on mental illness: primarily my own illness, but also that of the millions of other people who have suffered, are suffering, will suffer from mental illness. The changes that have transformed mental health care have been highly controversial. Some people applaud them as progressive alternatives to institutional care, while others condemn them as cost-cutting exercises masquerading as psychiatric modernization. This book does not adjudicate these disagreements (although it has things to say about them). I am not a historian of psychiatry, and anyway my own concerns are rather different. The rise of the Asylum Age witnessed the emergence of a paradigm conflict still evident in mental health circles. This conflict maps loosely on to the opposition between psychological and biomedical models of madness, but terms such as 'near' versus 'far', 'interactive' versus 'passive', 'subject' versus 'object', tell us more about it. How close to madness must a would-be healer get to effectively treat it? One paradigm shows us madness practitioners proffering remedies to sufferers across a gulf of professional expertise; the other shows sufferer and healer drawn together in a therapeutic partnership which, in the strongest versions of this tradition, directly implicates the healer's own psyche in the treatment process (an approach described by the leading forensic psychotherapist Dr Gwen Adshead as 'mak[ing] your mind available to somebody else, to help them recover'[1]). The demise of the asylum has reconfigured this conflict without diminishing it, and it continues to reverberate throughout psychiatric services. This book explores this conflict, not in order to resolve it but to show its importance for past and present understandings of mental illness and healing.

So *The Last Asylum* is a work of history – my own and many other people's. It is also a book about the work of turning the personal past into history. The lived past is never really past; it endures in us in more ways than we understand. Sometimes it doesn't even feel like the past; it just feels like life itself, like the way things have always been and always will be, now as before, then and for ever. For many people, experiencing the past *as* past, allowing it to become truly historical, is very difficult. It involves uncovering aspects of our lives, especially our early lives, that we have forgotten or have never really known about, except perhaps as occasional spasms of mind or body, disturbing dreams, strange movements in the blood . . . sequelae of things that have happened to us, or might have happened to us, or we imagine happened to us, hidden beneath our remembered pasts. The sources of self run deep. In my case, exploring them required me to become a historian of my own life, through the peculiar and demanding labour of a long-term psychoanalysis, of which *The Last Asylum* is also an account.

Psychoanalysis gets a bad press these days, especially as a treatment for severe mental illness. Most psychiatrists dismiss it, and it barely features in mental health literature. I was not textbook psychotic – even in my craziest times, I knew my hawk from my handsaw – but I was very ill indeed. Several psychiatrists whom I encountered during my years of institutional care urged me to give up my therapy. (One instructed me to return to Canada and marry a farmer, and became irate when I smiled at his prescription.) I was fortunate that my own psychiatrist was sympathetic to psychoanalysis and did her best to support me. Although she and my psychoanalyst had little contact with each other, they became for me a very effective working team. Without their partnership, and the many friends who looked after me during those years, I would probably not be alive today.

So this is, among other things, a narrative of gratitude – to the psychoanalytic process, and the analyst who practised it with me; to the many interesting people I met during my sojourn in the psychiatric world; to my friends and family; but above all to the crazy woman I once was, whose craziness in the end proved my salvation. This last point may seem strange. Madness as its own cure seems a paradox, and in a way it is. But the unconscious mind is a very paradoxical place. I am no romantic celebrant of lunacy, no 'madness is truth' prophet. There is much wrong with modern attitudes to mental illness – this book has things to say about that – but madness is a horrible state. To say that it has a creative dimension is not to soft-pedal its miseries. Modern medicine offers remedies – drugs, psycho-behavioural therapies – that bring relief to many sufferers; only someone who has never experienced mental illness would dismiss these. But madness is not a disease like any other. The crazy mind is no weak or failing instrument but intensely alive, bursting with inventive energy. This is a cliché of romantic ideology, but it is also true – and it presents a huge challenge to conventional medical concepts of treatment and cure.

What to do with creative unreason? The standard psychiatric strategies are management and suppression, primarily through psychotropic drugs. Wanton thoughts must be restrained; the unruly mind must be tamed. The alternative – to encourage the mind's extravagant energies, to draw its baroque vitality forward to its full extent – would seem true madness, and deeply irresponsible. Yet this is precisely the psychoanalytic project, which instead of suppressing the lunatic psyche engages with it, unleashing its dynamism inside the therapeutic relationship. Solo madness becomes shared madness, with one side of the relationship hopefully remaining sane enough to feel the craziness without getting lost in it. When psychoanalysis works, as it did for me, it is because the human mind is always trying so hard to comprehend itself and its world, but sometimes it needs help to do so. I was lucky enough to get such help.

This type of assistance can be hazardous. Psychoanalysis works through the transfer of the patient's illness on to the analytic relationship, but this famous 'transference' is a cool label for a very heated, sometimes incendiary, process. I came close to dying in the course of my analysis. I threw myself entirely on to my analyst's skill and inner resources, and the toll on him was often apparent. Yet psychoanalysis is far from unique in this. All close relationships trigger powerful emotions whose true origins lie elsewhere, often in childhood. 'Caring' relationships in particular evoke feelings of great intensity. Psychoanalysis is unusual not in inspiring such strong emotions, but in striving to understand them, and in taking responsibility for them. And when the feelings that are aroused are unbearable, as they often are in severe mental illness, this responsibility can be very onerous.

So gratitude is called for. Many kind people, especially

my psychoanalyst, put themselves on the line for me. I knew little kindness in my early years. This is true of many people, and their stories – 'misery memoirs' – fill our bookshops these days. Is this another misery memoir? Yes and no. Certainly there is plenty of misery in it, and not just my own; but what this emphatically is *not* is a tale of victimhood. The misery it describes is not of that order. There is no victimizer in my story, or if there is, I am it. Like most mad people, I was literally my own worst enemy: simultaneously victim and victimizer, brutalized and brutalizer. Every crazy person has a big investment in her misery. How else can she go on paying for her crimes? Mental illness is a war of self versus self – and anyone who gets between the assailants is likely to come in for a lot of flak. Many people around me did, and I am grateful to them for their tolerance and compassion under fire.

I set out to write this book thinking that I knew what I was doing. I had spent many years writing about women's lives, piecing together their stories from published writings, manuscript scribblings, scraps of this and that. Quite often when I was researching one of these women I would dream about travelling back in time, to ask her why she had taken that journey, fallen for that man, got embroiled in that controversy – the perennial fantasy of the historian. Now I had my primary source right here, seated at my desk, waiting to be interrogated. I also had documentation from my period of illness: a pile of journals, lots of letters, a few audiotapes. So I felt confident that the task was do-able; what I couldn't anticipate was its oddity ('What are you writing about now?' a friend would ask. 'Penis dreams,' I would say), or how big it would grow, as I harnessed my personal story to the history of the community care revolution, with its cast of many thousands. However, this larger historical dimension of the project also made it very sociable, as I met with other people who had lived through these changes and who generously agreed to share their memories with me. In this sense *The Last Asylum* has a collective dimension, which I hope is apparent in the text.

For the telling of my own history, I had expected to draw on the memories of close friends who, as the reader will see, played a huge part in my care and eventual recovery. In fact I did much less of this than I anticipated. The madness of a friend touches people in complicated ways. My friends have their own stories to tell about their relationship with me, about the experience of caring for someone who was too confused and wretched to care about them, about what it was like to witness a friend in deep crisis. 'I don't want to write your obituary, Barbara!' one of my closest friends said to me, at a time when it seemed quite possible that she would be called upon to do that. But the feelings that lay behind that *cri de coeur* were hers, not mine, and I am not entitled to tell her side of the story.

Some readers will wonder about the accuracy of my account of my illness. I don't pretend that everything happened exactly as I describe it. No historian can ever claim this, and sometimes I have felt the need to be tactful, towards myself as well as others. But I have been as truthful as possible. 'You want to know the truth about yourself,' my analyst told me early on, as if this might matter a lot to me. It did, but psychological truth is a shape-shifter, and conveying it is tricky. 'I hardly know how to write about myself,' the novelist Hilary Mantel says at the beginning of her memoir, *Giving Up the Ghost*. '. . . I will just go for it, I think to myself, I'll hold out my hands and say, *c'est moi*, get used to it.'[2] But the *moi* of my story wasn't present to me in this way. It was a self in process, a 'me' who came into existence in the years I describe here. So what I have to offer the reader are backward views from the vantage point of someone who both was and was not there at the time. Accurately remembered madness is oxymoronic; if you can *really* remember it, you are still mad. And the pain of some past experience pushes it below the radar. The writer and Holocaust survivor Charlotte Delbo could not remember her months in Auschwitz until the sudden moment, many years later, when she smelled the stink of the concentration camp on her skin.

Delbo describes this as 'deep memory' as opposed to 'thought memory'. Proust (who also had much to say about the mnemonics of smell) called such deep memories 'resurrections'. 'When these resurrections took place. . .,' he wrote, '[I] grappled like a wrestler with the present . . . If the present scene had not very quickly been victorious, I believe that I should have lost consciousness.'[3] So some memories are high-risk. I could not remember the experiences described here until I began to forget them, and the reader should keep this in mind when following me into my past.

My psychoanalyst – V as I call him here – played no part in the writing of this book. All psychoanalysts are, to a greater or lesser degree, fictive beings, creatures of their patients' needs and imaginations. This was certainly true of V, who at different times in my analysis was a god, a devil, a shaman, a snake-oil salesman. I never really met the man himself, apart from his analytic persona, nor will you in these pages. 'He seems to have been pretty tough on you,' a friend remarked after reading parts of this book in manuscript. 'He needed to be!' I replied, but I was surprised. My memories of V are mostly of his kindness and patience, but if that was all he had brought to our relationship I doubt I would have survived. Perhaps it was only in the writing that a truer picture could emerge. At any rate, the conversations between us that I recount here are true to my recollections of them – in substance if not in detail – but it is I alone who am responsible for recalling them in this way (this is also true of my conversations with my psychiatrist, Dr D). The psychoanalytic adventure is simultaneously shared and inexorably private. In the end, this is my version of my history, and nobody else's.

A note on language: the vocabulary of madness is controversial. 'Lunacy', long the official descriptor, fell from fashion many years ago, and most people today think that calling someone mad is unkind and stigmatizing and so prefer to speak of 'mental illness'. But is madness an illness? Most psychiatrists believe so, attributing it to glitches in brain chemistry. The fact that there is no compelling evidence for this has not prevented it from becoming a professional orthodoxy.[4] But of course 'illness' can mean more than physical ailments, and here I often speak of myself as ill (as well as mad). 'Ill-being' might have conveyed my condition better, but I have chosen to use the standard terminology of 'mental illness', 'mental disorders', et cetera, not because I endorse the biomedical view of madness implied by this choice of words, but because this is the nomenclature of madness in today's psychiatric world and *The Last Asylum* is about this world as it is, not as I might like it to be. However, like most polite descriptors, 'mental illness' has come to mean much the same thing as the terms it has replaced. It is striking that the most common psychological disorders – anxiety, compulsions, milder forms of depression – are seldom described as mental illnesses.[5]

So I talk about 'mental illness' and 'madness' here. I also, at different times, describe myself as 'crazy', or as feeling 'lunatic'. One of the fundamental premises of psychoanalysis is that we all – the officially sane as well as the occasionally mad – have a crazy substrate to our personalities, laid down in our earliest months of life. Most people can manage, and sometimes even exploit, this craziness without it taking over their inner worlds, but some cannot and later 'go crazy'. This does not make them nutters or schizos or whatever derogatory tag is currently in vogue. For the vast majority of people with mental illness, the illness is a now-and-then experience, *not* a fixed identity. (This is not even to touch on the complex issue of historical and cultural variations in what is deemed crazy: today's single-mum bisexual pop star would certainly have been regarded as asylum-fodder a century ago.)

Even in my worst times, the madness waxed and waned, until eventually I discovered its causes and made my peace with them. Today I am no crazier than I need to be – than we all need to be – to negotiate modern life. But the person I am, I became through my madness:

not by 'recovering' from it, which implies a return to a previously healthy state, but by entering into it and travelling to its roots. When I was a child I tried now and then to say something about my unhappiness. I was told that I exaggerated. I was a sulky, noisy little girl who made things up a lot, so I was sure that the accusation was true. The feelings I felt I didn't really feel; my sense of reality was flawed; people weren't as they seemed to be; I was not the miserable, frightened child I experienced myself as being. My madness lay there: in the unhappiness, which was in fact much greater than I knew, and in the guilt and confusion surrounding it, which it took many years and much effort to undo. I have made every effort not to 'exaggerate' here and the result, I suspect, is an account softened away from some of its harder edges. But some things are probably incommunicable except perhaps by poets, and writing (and reading) a book should not be an endurance test.

Mental illness is very ordinary. I needed to think of myself as extraordinary, for reasons that will become clear, and for a time I told myself that my craziness was exotic and special. This fantasy dissolved when I entered Friern. My years inside the psychiatric system taught me much and left me with many questions. When I departed the system I pushed these questions to the back of my mind. But they went on scratching at me, along with my friend Raphael's injunction to think historically about the old asylum where I had fetched up. So I knew from the start that this book would deal with the asylum closures. I had no background knowledge of this history, and spent over a year in the scholarly literature and the relevant archives. But my biggest insights came from the people I interviewed. I had long conversations with nurses and doctors who had worked in Friern, including one of my own nurses and my psychiatrist. I interviewed one former patient and tried to talk to more, but this didn't work out.

Friern was home to many people who were deemed incapable of managing in the community and so had to be found new dwellings when the hospital closed. Group residences with full-time staff were established for these patients, and some of these still exist. I went along to two of these residences to meet with people who had agreed to talk to me. In the first residence I met briefly with a woman who insisted she had never heard of Friern and was extremely anxious to give me a packet of crisps. She was cheerful and pleasant and I accepted the crisps and departed. In the second residence the people who had agreed to speak with me had changed their minds by the time I arrived. I left this place feeling very uncomfortable. I had spent nearly four years among people with serious mental disorders, and I had assumed this gave me the wherewithal, and the right, to chat with them about their memories of Friern. I decided I was wrong in this, and abandoned the attempt. I *did* interview other former asylum residents, but these were activists in the service-user movement who were accustomed to discussing their experiences. My conversations with them were moving and sobering. Most were pleased that the asylums had closed, but when I asked them for an assessment of the current situation the verdict was unanimous: 'Bloody awful.' I heard stories of neglect, of over-drugging, coercion and disregard for basic rights. 'We don't call it "community care",' one man said to me, 'maybe community indifference, or compulsion . . .' This book is not a study of the mental health system but inevitably, as I met people responsible for delivering this care, and people in receipt of it, I formed strong impressions and drew conclusions. What would happen to me now, were I a young woman in the midst of a severe emotional breakdown? It's a question I asked myself many times as I worked on *The Last Asylum*. The book concludes with the beginnings of an answer to this, in a survey of the state of mental health services in our post-asylum world.

PART ONE

I

Beginning

19 November 1977. I had stayed up too long and thought too much. That was what I told myself later. I had spent all day and all evening – as I spent nearly every day and evening – writing my doctoral thesis. At about eleven, as I was contemplating bed, I had a good idea, a fantastic idea, an idea so wonderful that it made all my previous ideas seem thin and silly. Excited, I tested it: would it hold up? I was sure that it would (it did – it became the basis of my first book), and my excitement became overwhelming. I lay awake all night, breathing fast, heart lurching. The sleepless night segued into a day of exhaustion and anxiety, followed by an endless stream of such nights and days. I had been an easy sleeper; now insomnia was my bedfellow, and my anxiety levels climbed steadily. A lightless misery engulfed me. The world drained of warmth and colour; a cold blankness was everywhere. This went on and on as – armed now with many good ideas – I continued to labour away on my thesis.

I had always been an occasional hair-puller, but now I yanked out hairs constantly as I worked. I pictured myself going bald, and began to wear a headscarf when I was writing. 'You put on your scarf and go grey,' a concerned friend remarked.

In 1980, as I crawled to the end of the dissertation, I landed my first academic job, teaching history in a small provincial college. Such jobs weren't easy to get, and I was very pleased. I took a room in the college town and started commuting, spending weekends in London.

It was a nice, friendly college – a good place to launch an academic career. But I could barely manage it. I dragged myself into classrooms, panting with fatigue and panic. Six months in, I attended the oral examination for my PhD, returned home triumphant, and collapsed. I lay in bed stupefied, my mind frozen in fathomless exhaustion, dark horrors moving beneath the ice. My throat closed up, my heart raced. Sleep, when it came, brought no relief. My GP, a medic of the pull-yourself-together school, prescribed sleeping pills that helped little. Then one dawn, after yet another endless night, I downed several of these pills with a big slug of whiskey. The exhaustion lifted. I had found my self-cure. Armed with booze and pills, I returned to work.

The conversion of intolerable emotions into bodily symptoms is one of the mind's favourite tricks. Freud would have called me a 'hysteric'. But this was 1981, when hysteria was long out of fashion, and before 'Chronic Fatigue Syndrome' had come on the scene. Had it been a few years later, I suspect my GP would have diagnosed CFS, or ME (Myalgic Encephalopathy) as it is also known in the UK. I can easily imagine myself settling into physical invalidism, or experimenting with the endless remedies on offer to sufferers from these conditions. Instead I turned to psychoanalysis – as much for its prestige-value, I suspected at the time, as from any optimism about its curative potential. The left-wing intelligentsia of 1980s London was infatuated with psychoanalysis. Hovering on the edge of this world, listening to people comparing analysts, swapping couch gossip, I yearned to join in . . . or so I thought. My 'illness' was surely just an alibi. I asked a friend who was well connected in therapeutic circles to advise me. She asked around, and eventually I found myself in front of Mrs B, an elderly and very distinguished psychoanalyst who had agreed to see me for a diagnostic consultation. Mrs B was a gentle-mannered woman who treated her patients in a beautiful summerhouse in her Hampstead garden. Sunshine flooded in through the windows, and

I remember sitting on the floor with my head in her lap. I know I did nothing of the sort but the memory persists.

> *I don't think you want interpretation. You will make a fool of your-*
> *self and you will hate that. It will be very painful for you.*
> *I don't know what to do.*
> *I don't think you have any choice, do you? But it will be very hard*
> *for you.*

I should see a man, Mrs B told me, and he should be the sort of analyst who cared more about his clinical work than writing clever articles for therapeutic journals. A fortnight later, she wrote to me to tell me that Dr V in Harley Street had agreed to see me. Harley Street! I was thrilled by the elite address, by the classiness of this new adventure.

V was in his mid-forties, which seemed about right – not old, but old enough to make me feel young enough to deserve looking after. A conventional suit; a neat Freudian beard. The Harley Street room was large and coolly hospitable: pale walls and carpet; leather sofa and chairs; a wooden desk in the style we used to call Scandinavian. A large pastel still life was propped against one wall; the top of the filing cabinet held a few ornaments that made my lip curl: china elves, pink ceramic figurines. None of it was to my taste, but nothing – for a while at least – really offended me. In the corner directly across from the door was an executive-style reclining chair, and in front of it a couch, *the* couch, which I ignored for eight months.

V was standing beside the recliner. He greeted me, and waited to see where I would go. I walked across the room to the leather sofa, as far from the couch as I could get without actually climbing out the window. He sat in a chair across from me, a coffee table between us, and we began to talk. The couch was not mentioned, but my eyes kept drifting towards it. (Eight months later I walked into the room, went straight to the couch, and lay down in the standard position, facing away from him. I enjoyed the look of surprise I caught before I lost sight of him. For years after he would ask me – why that morning? Why then? 'You tell me,' I would snap back. Then one day I had the answer: 'I knew you wouldn't eat me.' 'Yes, that was a good discovery.')

So, what has brought you here?

I rattle out my recent history. It doesn't seem to add up to much; I have been making a fuss about nothing. I wait for him to tell me this.

Sounds difficult. Sounds like you've really been struggling.

I don't like his sympathetic tone: is he mocking me?

Oh! Yes . . . I guess so . . .
Tell me about your background – your family and so on.

I had expected this; I had even rehearsed it a bit. I settled back and talked. He nodded occasionally, asked a few questions, commented on nothing. I talked some more. Once or twice he smiled at me, which made me nervous.

We will need to stop soon.

I am vaguely disappointed: is that it?

He tells me his fee and I tell him what Mrs B has said I should expect to pay, which is rather less. We talk money for a few minutes.

I spend too much on things.
What sorts of things?
Oh, I don't know . . . meals out, clothes . . . I spend too much on
stuff like that.
You need food and clothes. You work for your money; you should
have enough for what you need.

This remark disconcerts me, although I have no idea why.

Eventually we strike a deal: he will see me for the fee Mrs B had mentioned, so long as I agree to pay more when I start earning more.

OK. Yes, that's fine.
You sound quite sure you want to do this.
[Have I sounded sure? Am I sure?] *Do you think I should?*
Yes, I think I do.
Oh! OK then . . . yes, I mean, if that's OK with you . . .

So now I was an 'analysand' (how I loved that word!) and for a few months I buzzed with my new status. My twice-weekly sessions excited me profoundly. I was sexually charged, giddy with erotic aggression. Sitting across from V, I posed and preened and baited him. I felt like Kate Hepburn teasing Spencer Tracey; I felt gorgeous. To my friends I seemed wild, manic. 'Goodness, Barbara, what's the matter with you?' my friend Sally demanded, alarmed by my feverish talk and frantic energy.

The comedown from this was excruciating. All my earlier symptoms returned, much intensified. My exhaustion thickened to the point where it became difficult to speak clearly; the world was unfocused, mushy. Anxiety roiled around inside me, occasionally sparking into terror. I took more pills, began to drink heavily, staggered along. My delightful sessions turned rancorous, as I realized that my brilliant, charismatic analyst was in fact an incompetent oaf. He understood nothing; made stupidly inappropriate comments; wore prissy striped shirts with white collars. I arrived earlier and earlier for my sessions, trembling with anxiety that he might not be there, that my hostility might have killed him off. As my dependence increased, I became waspishly satirical, batting back every interpretation with snide retorts. My aggression terrified me, which only ratcheted it up further. During one particularly nasty session, after I had scornfully waved away a dream interpretation, V lost patience.

This is no game!
Don't shout at me!
I'm not shouting at you. I'm asking you to take yourself seriously.

To take myself seriously? Now *there* was a road to hell.

Before I went into analysis, I knew nothing of me. I lugged myself around like an incubus, an alien 'I-thing' that had been foisted upon me. All I knew about this so-called 'self' was that she was wrong and bad. Now I would find out why. After several years of analysis, when I finally began to meet myself, the person that I encountered was so angry she could destroy the world. To take her seriously meant entering a tumult of grief and despair so intense that my body folded up with it, so bloody with rage and hate that no one could survive it.

We can't go there.
We are already there.
Will I survive?
I believe that you and I and the analysis will all survive.

I had to run away, but he was all I had. I had to hang on to him; I had to give him whatever it took to keep him. But what did he want from me?

Perhaps he wanted what my parents wanted: books. This was what had always been wanted from me. I was born to write, to conjure words that the world would applaud; literary renown was my filial duty. The injunction lashed me to my typewriter, and booze and pills kept me there until I couldn't go on any more. Twelve months after beginning psychoanalysis I published *Eve and the New Jerusalem*, a reworking of my doctoral thesis on feminism and utopian socialism. The book was widely and rapturously reviewed and I was awarded the Isaac Deutscher Memorial Prize – the first woman to win it. I had no expectation of such success, and was stunned by it. My parents flew over from Canada for the celebration party. 'It's a terrific book,' a friend said to Mum. 'Yes, I'm sure,' Mum said, adjusting the strap on my dress. 'Her father and I are hoping the next one will be a novel.'

I gave V a copy of the book. He thanked me and told me he wouldn't read it 'until the analysis is finished'. Then I told him about the prize. He was silent. I felt a surge of what I thought was anger, but what I actually said was: 'Thank God you didn't congratulate me!'

Six months later I gave V another book that I had just published, a children's story. This he read, and a few days later he opened the session by complimenting me on it. 'Very nicely done. Perhaps you'll take up full-time writing . . .' After the session I went home and started writing. Then I began vomiting. All afternoon and evening I moved between my desk and the toilet, scribbling an outline of a new book which would be a great success, a bestseller, and which I would dedicate to V. When I lay on the couch the following morning and reported this project to him, I heard a sharp intake of breath.

What's the matter?
You are showing me how dangerous it is to praise you.
I thought you'd be pleased.
With what? The dedication, or the vomiting? You'll make me famous even if it kills you?

So what *did* he want from me? More time, that much I knew. He was always pressing me to increase my sessions, from two a week at the beginning of the analysis, up to three a week the following year, four during the college vacations. Between teaching and commuting, I couldn't fit in any more. Then I received a one-year research grant, to write a book on the feminist pioneer Mary Wollstonecraft. This meant that I could stop teaching and start a full-time analysis. To be in psychoanalysis 'full-time' means a session every weekday, but in my case it meant literally all the time. I told myself that I could combine analysis with writing my book; but in the autumn of 1984, as I trudged to my sessions, psychoanalysis took over my life. Soon I could think about nothing else. At the end of the year, as my funding dried up, I resigned my job, and threw myself on my parents' charity.

They never failed me financially, but before I could take this step I needed another guarantee.

> *Will you go on seeing me, even if I don't have the money to pay you?*
> *What, not pay me at all?*
> *No, I mean not like now, not proper fees.*
> *Hmm* . . . [Long pause] *Yes, yes, I will. But I shall expect you to pay whatever you can afford.*
> *Oh* . . . *yes, OK.* [Long silence, then tears] *I thought you might stop, you know, if I ran out of money.*
> *Just stop? Just like that?*
> *Well* . . . *you know, if I couldn't pay* . . .
> *We have gone too far for that.*
> *Yes* . . . [Suddenly an astonishing thought occurs to me.] *You want to analyse me!*
> *Hmm* . . . *that's your trump card, isn't it?*

I dance away from this session. He loves me! He loves me!

I haven't even made it back to my car before the contempt hits me – the stupid fool, the idiot, to think he can analyse me! – followed by a spasm of terror: how will I ever get free of him now? What will he do to me?

2

History

My friend Cora gave me a hardback notebook. I had been in analysis for a couple of months, talking about it incessantly. 'I thought you might like to write some of this down.' That evening I made my first entry. When I came to the end of the notebook I bought another. By the time I left analysis, in 2003, I had thirteen notebooks, plus various unbound scribblings. 'An archive,' Cora said when she saw the stack of notebooks.

There was no pattern to my journal-keeping. Sometimes a month or more would go by with no entries, then I would write every day for a week or so, occasionally two or three times a day. Some of these entries were long and reflective; some were written when I was drunk. I wrote mostly about V and me, but also about my friends, my family, childhood memories. I recorded many dreams. At first the writing was difficult – 'I've never talked to myself before' – but soon the words came easily. 'No one reads this except me,' I told myself. I didn't tell V about it for nearly a year.

You were afraid I would ask to read it.
No, no . . . it just didn't seem important.
And what if I asked to read it?
You wouldn't!
Are you sure? I would probably learn a lot from it. You're a historian,
    after all.

What's history got to do with it? [I hear the silliness of this even as I say it.]
Everything.

One of my earliest journal entries was about a book I had loved as a child. This was a catalogue from a photographic exhibition titled 'The Family of Man', mounted by the Museum of Modern Art in New York in 1955. The exhibition, a visual depiction of the human life cycle in societies around the globe, was the most popular in MoMA's history; people travelled from across North America to see it. My parents didn't get to the exhibition but a New York friend sent them the catalogue, which I was allowed to peruse once my hands had been checked for cleanliness.

Here was a beautiful object! Five hundred gorgeous photos by top photographers, interspersed with ancient proverbs, Old Testament verses, epigrams from famous writers, all celebrating what the left-wing poet Carl Sandburg, in the Prologue, called the 'holy kinship' of humankind:

> The first cry of a newborn baby in Chicago or Zamboango, in Amsterdam or Rangoon, has the same pitch and key, each saying 'I am! I have come through! I belong! I am a member of the Family!' People! Flung far and wide, born into toil, struggle, blood and dreams, one big family hugging close to the ball of earth for its life and being. People, the living flowing breath of the history of nations . . .

I was five when I first saw this book, and very soon it became my bible. I spent many hours and years poring over its photos of Indian mothers and babies, African-American lovers, Nepalese shepherds, Nigerian children at play. After I had learned to read I studied the proverbs and quotations, and spent a lot of time pondering a full-page spread showing the newly formed United Nations in session, one side of the image overlaid with the opening lines from the UN Charter declaring its members' determination to 'save succeeding generations from the scourge of war' by reaffirming the 'dignity and worth of the human person'. The faces of the U N delegates fascinated me, so varied yet charged with common purpose. Here, a mere decade after V E Day, after the bombing of Hiroshima and the liberation of Auschwitz, was a stunning depiction of world unity, a 'camera testament' in Sandburg's phrase, to the 'essential oneness' of humankind in the wake of total war. I didn't get the political message until much later on, but I memorized many of the proverbs and Carl Sandburg became my first literary idol.

In the years after its New York opening, 'The Family of Man' exhibition went global, travelling to thirty-eight countries, where it was seen by over nine million people. One of its first stopping points was Paris, where it was viewed by the cultural critic Roland Barthes, who promptly wrote a scorching indictment of it. Where, Barthes wanted to know, was 'the determining weight of History' in this 'myth of the human "community" '? Where were the historic realities of injustice and oppression in this celebration of the natural kinship of humankind? '[L]et us . . . ask the North African workers of the Goutte D'Or district in Paris what they think of "The Great Family of Man".'[1]

I read Barthes's essay in my mid-twenties, and it shocked me. I had grown up on these moving images; in fact as a very little girl, accustomed to family albums of the usual kind, I had believed that some of these pictures *were* family photos. Here was my mother, in bed with my baby sister and me; here my sister in her crib; here me, with the family cat (we didn't have a cat at the time). The illusion didn't last long but the photos retained their emotional charge. Moreover, I was a child of the political left, a 'red-diaper baby', raised on homiletics about the fellowship of man. The romantic humanism of 'The Family of Man' was my parents' creed; who was Barthes to gainsay it?

And I knew all about the history of injustice and oppression — Barthes had nothing on me there. In fact I had been raised on just such a history. 'History begins,' Vladimir Nabokov wrote in his memoir *Speak, Memory*, 'with a grey engraving in an ebony frame –'

> one of those Napoleonic-battle pictures in which . . . one sees, all grouped together on the same plane of vision, a wounded drummer, a dead horse, trophies, one soldier about to bayonet another, and the invulnerable emperor posing with his generals amid the frozen fray.[2]

This picture had hung in the four-year-old Vladimir's home, above a 'big cretonne-covered sofa' where he had liked to play. His father had served in the Horse Guards, and the picture became for Nabokov one of those emblematic family objects that like 'an anonymous roller' 'pressed upon my life a certain intricate watermark whose unique design becomes visible when the lamp of art is made to shine through life's foolscap'.

For me also History began with a war picture hanging in the family home. In my case this was a big framed print of Picasso's painting of the 1937 bombing of the Basque town of Guernica, which occupied pride of place in our living room. I cannot remember a time when I did not know every inch of that horrifying painting which for me, as for Nabokov, showed History as paternal valour: Dad had fought with the International Brigade in the Spanish Civil War, an experience which overshadowed everything that followed, including his marriage to a fiery young radical eight years his junior. 'Your father,' Mum would say to me, 'was there – right there! It was terrible, terrible . . . but everybody looked up to him. He was a real leader.' I grow up on endless stories about this and later displays of Dad's left-wing courage. Mum trumpets her own battles too, but less, or less overtly, yet over time I come to understand that she also is a brave freedom fighter. I am a child of History in the heroic register; I am a daughter of the greatest parents ever; I am the luckiest girl in the world.

There is a passage somewhere in a Penelope Lively novel where she suggests that children are repositories of parental moral commitments long since forgotten or flouted by parents themselves. Whether or not this is true, I know that my parents' tales of radical valour took root and grew in me until they became history itself.

We lived in Saskatoon, a small city on the Canadian prairies. Mum was the oldest daughter of a Jewish middle-class couple, pious and conventional. Dad was an emigrant Welshman, scion of a Communist mining family. Their marriage, when Mum was barely eighteen, was a family scandal, especially when Mum joined Dad in the Communist Party. It was 1940, and sides were being taken everywhere. The element of truth in my parents' historical romance lay here, in the political heat of the times that gilded their past with a radical-heroic lustre which never palled in their eyes, even when their adult lives became settled and respectable. In the late 1940s Dad trained for the law and joined my maternal grandfather's law firm. My mother then also trained as a lawyer. In 1955 I watched her graduate. In 1960 she became a magistrate, later a judge.

Dad and Mum both resigned from the Communist Party in 1948 (over Yugoslavia's expulsion from the Comintern) but they remained on the political left. Dad was a labour lawyer and an activist in Canada's left-wing party, the Co-operative Commonwealth Federation, now the New Democratic Party. He ran for Parliament twice, unsuccessfully; tried to become leader of the provincial NDP, also unsuccessfully; and then settled into the councils of City Hall, where he battled with local Babbits and became a popular spokesman for Saskatoon's working-class population.

My mother's position on the bench prevented her taking any direct part in politics. But her views too remained staunchly left wing. Mum was keen on Chinese Communism; copies of *China Reconstructs*, a glossy magazine crammed with beaming peasants and warm-faced functionaries, littered our house. She was a prison reformer, peace campaigner, and a passionate supporter of the civil rights movement. In 1972, when she was fifty, she was made Chairwoman of the Saskatchewan Human Rights Commission, the first in Canada, and went on to mount a series of increasingly controversial campaigns for women's rights, especially the right to abortion, and native Canadians' rights. At the end of the 1970s she published a report by the Commission accusing a provincial government department of racist corruption. She was isolated, pilloried, and eventually forced to resign.

These are estimable histories, deserving of applause. For thirty years, they blanketed my moral horizon. Being Dad and Mum's daughter was the most important, the most worthwhile, the only worthwhile, thing about me. My reverence for them occasionally lifted a friend's eyebrow, but I paid no heed. Nor did the gulf between their political values and their personal behaviour disconcert me in the slightest. Dad was a domestic tyrant of the old school, Mum his adoring slave. As an adolescent I had a stormy relationship with Dad – but so what? Kids were a pain; I was a pain . . . Dad was just 'old-fashioned', and I didn't deserve him. I would shake my head over my selfish silliness.

In my early thirties, before I became too unwell to travel, I went back to my parents' home for a visit. While I was there, I fished *The Family of Man* out of the bookshelf. I stared at the familiar photos, disconcerted. Here were all the beautiful images I remembered, of people working, playing, love-making; but here also were others, very different – a crowd of Korean women screaming behind barbed wire; a gang of starving Indian beggars; Jews being herded down a Warsaw street by German soldiers. And children: a terrified girl roped to a tree; a young boy raising a heavy stick against a screaming woman. One page showed five photos of small children, alone, frightened, one of them a tiny toddler trailing behind a woman (his mother?) who is ignoring him, arranged around a few lines from the Southern writer and civil rights campaigner Lillian Smith: '. . . deep inside, / in that silent place / where a child's fears crouch . . .' Looking at these photos again I recalled them all, but now my memories of them sharpened, took on a razor edge.

One photo showed a solitary girl huddled on a bench in front of a blank wall. The girl's knees were pulled up, arms wrapped around them, face buried in her arms. Underneath her picture was a line from the fourteenth-century Chinese poet Lui Chiu: 'I am alone with the beating of my heart . . .' I remembered how as a child I had fretted over this girl: what had happened to her? The room in which she sat was empty. I used to think maybe it was a waiting room in a deserted bus station and that whomever the girl was waiting for had not shown up, would never show up. Or perhaps the other person had just left, never to return. Carl Sandburg had written that anyone looking at the human emotions on display in 'The Family of Man' was likely to say to themselves, 'I am a member of the Family. I am not a stranger here.' But what about this girl? What kind of lonely stranger was she?

Without history, there is only noisy silence. If we expunge history from the human story, Barthes wrote, 'there is nothing more to be said'.

The first time I read Barthes I was busy turning myself into a historian. This was not something I had intended. I didn't study history in school, and the only history book I read before I was twenty was a propaganda text for the Cuban Revolution. But in 1973, when I was beginning a very dull doctoral thesis on the philosopher John Dewey, I went to a women's history conference, one of the first held in Britain, and I was hooked. I had been active in Women's Liberation in Canada and now I saw an opportunity to connect my feminism to my academic work. Within a month I had ditched Dewey for a thesis on the history of early socialist feminism. It was the heyday of History Workshop, a network of academics, students and community-based historians working to build a radical-democratic 'people's history'. Women's history in Britain grew up under the History Workshop banner, and I soon became an active History Workshopper (as I remain, although the larger network is long defunct).

The book that emerged from my PhD thesis, *Eve and the New Jerusalem*, was a historical defence of the values underpinning 'The

Family of Man' exhibition. (I had no notion of this at the time, but it is obvious to me now.) *Eve* was a study of the first British socialist movement, the Owenite movement, as it was known after its founder, Robert Owen. The Owenites were militant exponents of family values socialist-style. Their socialist ideal was a world bound together by quasi-familial ties, governed by love and sympathy instead of competition and aggression. But there was a complicating element in this vision. The Owenites were, in theory at least, sexual egalitarians, committed to equality for women in every sphere of life. This feminism set them apart from other radicalisms of the period. But the movement's glorification of familial unity worked against their feminist principles. Family life – as Barthes might have pointed out, but didn't – is hardly a model for equalitarian living. The idealization of family is not just anti-historical in Barthes's sense; it serves as a cover-story for very specific forms of oppression. The power of men over women, the power of parents over children: these aspects of family life were insufficiently acknowledged by early socialists (or later ones, for that matter). I knew this, and wrote about it in my book, yet I clung to the Owenites' familial utopianism . . . and dedicated *Eve* to my parents and sister. To me, history was an exercise in family allegiance.

I completed *Eve* a month after I began psychoanalysis. If Barthes's call for historical truth-telling had once offended me, this was as nothing compared to the fury with which I now greeted my analyst's efforts to probe my happy-family story. I wanted to change but I needed my stories to remain the same. My stories were me; they were all I had that was good. Without them I was merely bad; without them I was nothing.

How had I become bad? This too was a historical question, a Barthesian question, and so beside the point. My badness was eternal. Like birth, like breath, it was a condition of life. I was bad the way God is good. My badness was like the stink of shit; it was intrinsic to me. 'That's just Barb,' Mum would say, and I would hear her resigned disgust. Occasionally she would add that I was selfish, or pushy. And I was very greedy: this I was told repeatedly. Once, when I was about ten or eleven, Mum told me that I thought too well of people and then changed my mind. I had just met a girl whom I liked so this criticism made me anxious. It also puzzled me. Now I can see what Mum meant, that I wasn't good at ambivalence – which was certainly true, but then neither was she. All her criticisms of me, without exception, were accusations she made against herself. I was merely an extrusion from her, like snot or shit. I was she, and she wanted to love me and could not. She was me, and I loved her to death. We were locked together in a death grip, Mum and me. I was strapped to her like a suicide bomber; the smallest move on my part, and we were both a goner.

Dad was our lover man. Mum liked to tell a story about taking me on a car journey when I was four. I insisted on sitting on the front seat next to Dad, '*I'm* Mummy now,' I announced. Such an ordinary story. She thought it was funny and maybe for most children it would be. But I *was* Mummy, and as I got older the war over Daddy became a nightmare of envy and hate.

This was the history buried behind my family story. Such things are as common as muck, they happen all the time. People go crazy inside such relationships all the time. Mum and Dad did their best for me; I know this now and I grieve for it. Nor do I assume that their failings are the whole story. Perhaps I was a particularly difficult or fragile child? Maybe an easier or tougher kid would not have broken down? Like most historians, I am not keen on counterfactuals. The story I tell here is about my early life as I rediscovered it during the first decade of my psychoanalysis, as I travelled into what Barthes called the 'ulterior zone' of human experience, to turn a mythologized past into a credible history and a liveable present.

3

Spain

The bomb fell just moments ago. Where is George? He is not visible but he is there. Just there, huddled behind that wall, peering over at the screaming mother cradling her broken baby. Or maybe over there, crouched under the window from which a detached head balloons, its eyes creasing in terror. I can feel George's helplessness, the heavy thud of his heart as he listens to the shrieks of the dying horse and smells the blood. The pounding in his ears is so loud that he does not hear the bomber's drone. His chest constricts, all breath goes. He watches, and falls, and disappears – and reappears each time I turn back to the impossible scene.

The Basque town of Guernica was bombed on 26 April 1937. A few months later Dad travelled from Canada to Spain to defend the Republican government against General Franco's insurgency. George was twenty-two. A clip from the local newspaper, 'Saskatoon Boy Serves Machine Guns in Spain', shows him on his way to join the Mackenzie-Papineau Battalion of the International Brigade. Poorly equipped and barely trained, he saw sporadic action for six months before finding himself on the banks of the River Ebro. Halfway through the savage four-month battle there, which all but annihilated the Republican forces, the Spanish government withdrew the *brigadistas*. But Dad was already out of action, hospitalized for pleurisy in Barcelona. He never breathed easily again. Returning to Canada in 1939, he travelled across the country describing his experiences and calling on his compatriots to join the fight against Fascism. 'I may have been discharged from the Spanish army,' he told reporters, 'but I have not been discharged from the cause for which I want to fight. Wherever I am, I will continue to fight for democracy.'

I grow up on this story. Dad is a revolutionary hero. Handsome, charismatic, his blue eyes burn with love of humanity. Women are wild for him. Mum can't believe her luck. Sitting together on my bed, when I'm thirteen, she tells me again about meeting him at the Canadian Youth Congress in the autumn of 1939, and how he came up to her afterwards and put his hands on her shoulders. What then? I squirm with a feeling I don't like. 'He was there to talk about Spain. Everyone went to hear him.' These days he doesn't talk about Spain, or at least not to his daughters. He leaves that to Mum, his adoring child-bride, who talks about him incessantly. I am meant to hear what she heard, feel what she felt. I don't demur; I know Dad is superman, *numero uno*; yet I am never, it seems, quite appreciative enough. My mother compares him to Marlon Brando and Richard Burton – 'real men', she tells me, 'not like those Beatles'. It is 1963 and my sister and I are Beatles fans.

Dad's radicalism was a family inheritance. Both his parents were dedicated Communists. Grandad mined coal in the South Wales valleys. He was a fierce proletarian warrior, savage in his loathing of the pit owners and bitter in his denunciations of class traitors like the union officials who sold out the miners during the 1926 General Strike. 'Bosses' men!' he would yell, thumping his fist on the table. 'Bosses' toadies!' Migration to Canada in 1929 in no way dampened Grandad's fires, nor Grandma's, a former domestic servant who soon after arriving in Canada immersed herself in the unemployed workers' movement. In 1941 the Canadian government rounded up 'enemy aliens' for internment, and took the opportunity to imprison as many Communists as they could track down. Grandad was interned alongside German and Italian Fascists. Grandma had her relief payments cut off and was forced to live on charity until the Soviet Union joined the Allies, and Grandad was released.

They were a formidable couple, my grandparents, rich in certainties, prepared to risk all for the socialist cause. And they expected no less from George, the oldest of their three boys. From birth, Dad was marked out for great things. As a child, he did brilliantly at school, winning every scholarship in sight. But such childish achievements meant little to Grandad and Grandma as they waited for History to claim him. I doubt he had much choice about going to Spain. Many years later, after I have broken down, he tells me he believes that the pleurisy that pulled him off the battlefront and nearly killed him had been an escape ploy, a physical flight from insupportable fear. 'So I understand your illness,' he says. I am astonished by this remark, but when I ask him to say more he waves me away.

Dad is angry much of the time. As little girls, Lise and I try to keep out of his way. As I reach adolescence, he and I begin to fight. Soon these battles become almost daily occurrences. We are talking; a statement is made; the tone changes. Dad stiffens; his cheeks darken. My heart rate quickens. Assertion is met by counter-assertion; our voices rise, rise higher, smash against each other. Dad is shouting at me, his face red and contorted; I am screaming, weeping. A climax is reached. 'Enough!' he yells. 'Get out of here!' Or I have already left, hurtling up the stairs to my bedroom. An interval passes and usually nothing more is said, at least not by us; my mother scolds me: 'You don't know what you're talking about.' Occasionally Dad demands an apology. I know he is right, he is always right, so I give it willingly.

Dad shouts at Mum all the time, over missing socks, lukewarm tea, overcooked meat. Mum is a judge. She sits in court, robed in black, then comes home to his tantrums. They spend a half-hour in bed every lunchtime, the door shut.

Like Dad, Mum is a wonder of nature. I know the stories by heart: how as a tiny tot Tillie was hoisted on to the dining table to recite bits of her original verse; how all the neighbours marvelled at her precocity. She began attending university at fifteen; the following year she became Dad's lover and joined him in the Communist Party. At seventeen she announced to her parents that she and Dad were to be wed.

My maternal grandparents were Saskatoon notables, leaders of its small Jewish community. The prospect of their brilliant daughter marrying a Communist goy was intolerable. I never knew my Baba, who died when I was three, and barely knew my Sede, whom I saw rarely and who seldom spoke to me. Sede was a Rumanian who had arrived in Canada as a teenager. A successful lawyer, pillar of the town synagogue, he was a handsome, elegantly dressed man with a reserved demeanour. I listened to my mother complaining about his starchy manner, his conservative cultural attitudes, his Zionism (which she loathed); but to me he was just a cipher, a Mitteleuropean patriarch from a world for which I cared nothing. The Holocaust, from which he had rescued several family members, framed Sede's life, but I didn't want to know anything about this (and worked hard at not knowing anything about it until my thirties). What I did know, in the way that a child comes to know, was how much Mum yearned for Sede's approval, and how she suffered when this eluded her.

Mum once showed me a letter that Sede had written to her in 1941, a few months before she married Dad. 'Take a look at this, Barb! How's this for a father's wedding letter to his daughter?' Sede writes to her in a panic of rage. He is begging her, threatening her, warning her that such a wedding will probably kill him ('any excitement causes palpitations of my heart which might prove very serious'). She is not just an ordinary girl, he reminds her; she is the daughter of 'one of the most prominent Jews in western Canada', a piece of good fortune which requires her 'to maintain certain standards', and which could open doors that will 'slam shut' if she marries this Communist goy. Marry George, and everybody will hate her, he tells her; she will become a *Meshamedes*, a traitor to her people; her entire family will be ostracized. And for what? To satisfy excessive lust. 'Mother and I have read every page of one of your diaries,' he tells her, concluding that she has a 'sex problem'. She is obsessed with sex. 'Every one of your thoughts ends up with sex.' He doesn't blame her for this; she was just born 'over-sexed'. 'You cannot help it, but you must control it.' He signs off with a little joke about a dog who chases his tail on a railway track and gets its head cut off by an approaching locomotive. He makes the joke's violent lewdness explicit: 'many a pup loses its head running after a piece of tail'.

Mum watches my face as I read this letter. I am not as shocked as I ought to be. I know, without being told, about the kind of parenting Mum had. I feel, without knowing that I feel, the coldness that has seeped down through the generations. Baba, Mum once said to me matter-of-factly, 'didn't like hugging, and that sort of thing'. What Baba liked, what Sede liked, what mattered above all else to them, were Jewish tradition and public reputation. Children existed to maintain the first and to burnish the second. This conviction brought them closer to Dad's parents than either set of grandparents would have liked. 'They cared a lot about what people thought of them,' Dad's uncle, my Great-Uncle Vic, said to me one day. 'And George was the boy to make the world sit up. He had to be the best. Always the best. They were very hard on him, quite cruel really.' Telling me this embarrasses him, but it is no surprise to me; I have grown up with this terrorized prodigy. How did such a wonder boy ever find a woman worthy of him?

Quite easily, as it turned out. Children who have been mistreated in this complex fashion often turn to each other. All Dad's life, even when he was well established and prosperous, he behaved like a poor lad on the make. Meeting political notables, he would swell and stumble with anxious excitement. Mum was there to bolster him, to remind him who was the best. No one ever did this for her. As a child of class privilege she was more socially adroit than Dad but equally insecure, and possibly even needier of public acclaim. Even in late old age, after her mind had been devastated by a cerebral haemorrhage, the first

24

thing Mum wanted to know about someone was how 'important' they were. She debated with herself whether she had been more important than Dad or vice versa; she sighed over lost importance; she told people how important I was. Once I asked her what 'important' meant but got no response, presumably because to her this was so obvious it wasn't worth stating.

To Dad and Mum, being important meant living beyond oneself, in the admiration – and the fear, since they both liked people to be afraid of them – of others. Life lived inside an uncelebrated self, beyond the public eye, was a non-existence, a nullity. To be unimportant, to be *ordinary* – Mum would spit the word out like a curse – was a living death. People were either important or better off dead. Moreover, death by insignificance was no meaningless concept, since other people – who for Mum and Dad existed mostly as potentially appreciative but also dangerously envious self-projections – were always wanting to kill you off. You had to be on the alert for this, ready to stave off a world that could destroy you by demonstrating that the person you were just wasn't important enough.

Inside this drama of idealization, rivalry and envy, family life became a black hole of the mundane, waiting to suck you into annihilating oblivion. Mum felt this danger particularly strongly, having been raised to cover her parents in glory while at the same time enacting her role as a good Jewish hausfrau: a near-impossible feat that she nearly pulled off. What she lost in the process was her private happiness, sacrificed to a life spent refilling the bottomless bucket of public reputation, her principal source of self-esteem.

As a child and adolescent, I understood nothing of this. The miasma of grandiosity that filled our home clouded everything. Everyone talked so that no one could hear anything. Nonetheless, some messages got through to me with crystal clarity: I was a bad person; I must become a successful person; to become properly successful, I must write books. The last of these was an absolute. To be a celebrated writer was the imperative, the great achievement of which every other feat fell short. Both my parents attempted it. Dad wrote a novel about wartime Spain; Mum wrote poetry, short stories, a book about a disabled cousin. None of these writings were much good, which embarrassed me when I realized it. Yet Mum went on striving, and even in very old age still spoke of authorship as her great ambition. 'You are more important than me, Barb,' she would say to me wistfully; 'you are a real writer.'

April 1983. Tortosa, Spain

8 a.m. I am stretched out on the hotel bed, gasping for air. The white walls of the room gleam with morning light, and there are movements in the corridor. Mum and Dad are in the room next door, and soon one of them will knock on my door. I should get up . . . But I cannot move, I am trapped inside my dream.

We are on the hillside above the Ebro delta. Mum and I are propped against the car while Dad moves around with his camera. The sun is savage, and I am wretched with heat and dirt. We seem to have been travelling for ever. I have no idea where we are staying. I am ferociously hungry. I cannot take much more. I see a roast chicken in the back of the car, fat and golden, and become desperate for it. I reach for it, but Mum intercedes. She is furious, eyes popping with rage. 'What's wrong with you?' she screams, and grabs at me. Dad hears her and turns and makes a run at me, waving a gun. The chicken, I now see, is a baby, and I am tearing at it. 'Shoot her!' Mum shouts, and I duck. But it is the baby Dad shoots. Its head bursts and falls back; bloody stuffing sprays across the car seat.

My heart stops; and Mum is at the door. 'Barb, Barb, are you awake?'

*

26

Monday of the following week. I am with V, recounting my dream.

It was horrible, really horrible.
Yes.
Why a chicken? A roast chicken, for God's sake!
A golden roast chicken, you said.
Yes, so what?
What is golden?
I don't know — sunshine, jewellery, money . . .
Hmm . . .
Golden, gold . . . oh, it's my name . . .

I am Barbara Gold Taylor (after my mother's maiden name, Gold-enberg). Still, why a chicken?

This 1983 journey into Dad's wartime past is not his first. He has been to Spain with a group of vets a few years earlier — old *brigadistas* from Canada, the US, Britain, bussing around from battlefield to battle-field. Dad, who has nervous bowels and hates buses, had found this trip difficult. A family expedition, with him in command, would be much easier.

So this trip with Mum and me has begun in a luxury hotel in Bar-celona. We stroll through the Old Town and Dad shows me the hospital where he was taken after he collapsed in a Barcelona cinema while on weekend leave, his lungs filled with fluid. We walk from there into the nearby street where he convalesced. He points to a building further down the street. 'Maria lived there.' Maria is the girlfriend whom I used to hear mentioned now and then during my childhood. 'You would have married her, if the Republic had won,' Mum would tease Dad. 'I have Franco to thank for my marriage.' After the fall of the Republic, Maria had fled to a refugee camp in France, from where she had written to Dad — cheerful, affectionate letters telling him about conditions in the camp ('The food is splendid and I am a little fatter . . . But is difficult to get soap so it's better to have short hair') and twitting him about neglecting her ('I don't know why you keep silent in response to my letters. Is it because you don't want anything to do with Catalan women any more?'). After Dad goes into a nursing home in 1999, I find a few of these letters, which a Spanish friend translates for me. 'I will probably move to another country,' Maria writes, 'maybe Mexico or the Soviet Union. When I know more precisely, I'll write to you. Take good care, George, I would like to see you healthy and strong again.' This was her last letter; Dad had no more news of Maria, had no idea where she ended up. 'She was lovely,' Dad tells me as we look up at her window. 'I wish I had seen her again.'

In Tortosa we stay in a gorgeous converted fortress, and on our first night there Dad and I argue about Ireland. I am in love with an Irish American named Jack. Jack ought to be a man after Dad's heart. Small and cocky, he is a real class warrior, a proletarian intellectual steeped in left-wing lore, with a strong appetite for confrontation. I find him extremely sexy, with his truck-driver swagger and Celtic good looks. He is smart, funny, sentimental, seething with literary ambition. My parents meet him in London and Dad immediately takes against him. 'Trotskyist, is he?' To be a 'Trot' is diabolic. 'No, of course not!' I say, uncertainly (in fact he is, but I'm not politically savvy enough to recognize this).

Like most Irish Americans, Jack is passionately attached to Ireland. Dad, the ardent ex-Welshman, detests the place. Now, sitting around a hotel table on a hot Catalonian night, Dad and I quarrel about the IRA. I am missing Jack terribly; in fact, I am frantic with my need for him, haunted by a horrible conviction that when I return to London I will find him gone. Devastating loss feels imminent; my muscles are tight with fear. Yet when Dad throws down the gauntlet – 'Irish republicans are nothing but Catholic bully-boys' – I cannot resist. My understanding of Irish politics is weak, and Dad's is little better; but that doesn't stop us from loudly quarrelling until Mum, conscious of the other diners, hushes us. Back in my room I pace and storm, hot-chested with humiliation and obscurely angry with Jack, who ought to be there to support me. 'A real Stalinist, your daddy,' Jack says when I tell him this story, which further annoys me.

Later, I stand at the window, my forehead pressed hard against the darkness, overwhelmed with anxiety. My parents are right there, just yards away; it is unbearable. Leaving me at my door, Mum runs her hand through my hair. 'Kiss me goodnight, dear.' Her cheek smells of the sun, and I pant for escape. 'This trip is so important to your father. I am so glad you could come with us.'

I go to bed and dream of dead chicken-babies. The next day I can hardly move. I take tranquillizers, plead a headache and lie all day in the hot room, my heart drilling through my chest like a jackhammer. What is happening to me?

What was I so frightened of?
You had no idea?
No.
We shall have to find out.

This reply infuriates me. I have been in analysis for a year now; why can't he just *tell* me? It will be another eight years before I begin to understand how psychoanalysis works.

4

Genius

I never intended to leave Canada for good. Why would anyone want to leave Canada? A couple of men I encountered in my early years in London asked me to marry them so they could become Canadian. As a child I was repeatedly told how nice Canadians are, much nicer than Americans (the only Americans I knew were very nice, but they were clearly exceptions). Travelling in France in 1963, Mum pinned little Canadian flags onto Lise and me. 'We don't want anyone thinking you're American.' People who – hearing our accents and not noticing the pins – assumed we were American would apologise when corrected (they still do). So I never planned to emigrate and was not pleased – some three decades on – to realise that I had.

But Britain was the 'old country', the 'mother country'; Dad's parents always referred to it this way. The soubriquets should have warned me. Travelling to London in 1971, I was journeying to the starting place, the home place. And I needed a home. So my year of postgraduate study at the London School of Economics stretched into two years, as I undertook my doctoral thesis, and then into many more years as I struggled to complete it. By the time the thesis was finished, I was breaking down. So no return to Canada then, and none for the twenty-one years of my psychoanalysis. By the time the analysis was finished I was partnered and ensconced in a new family with deep London roots. So that was that.

People belong where they are born, amidst the images and smells and accents of childhood. Emigrants are haunted people, carrying inside them the ghostly landscapes of their origins. Many people have a desperate need to emigrate, as I did, but even when emigration is a welcome escape it usually involves deep loss. 'I'll never live in Canada again,' I thought on a flight back to London in 2001 and was engulfed by sadness. My partner looked at me quizzically. 'No going home again,' I said to her. 'No' she agreed. 'No going back.'

Christmas 1983. Saskatoon

I am visiting with Mum and Dad, staying in the house where I grew up. *Eve and the New Jerusalem* has just been awarded the Isaac Deutscher Memorial Prize. Just before leaving London I had received a letter from an Oxford college, informing me that I was long-listed for a fellowship there. When I tell Mum about the fellowship, she responds by telling me how as a toddler I once spoke to her in rhyme. 'So I thought, why shouldn't I have given birth to a genius?' 'Just like you,' I reply, then – vengefully – 'that's why I'm so ill.' Later she repeats this conversation to Dad, leaving out the bit about illness.

I have brought with me a new book, a volume of women's writings about their fathers, which includes a poem by me titled 'Freud: Father'. The poem uses phrases and images from Freud's case history of the neurotic adolescent Dora to express some of my feelings for Dad. It is vaguely erotic, with a violent undertow. Presenting Mum and Dad with a copy of the book, I tell them that the poem is about V, not Dad (which, reading it again all these years later, I see is probably true).

Dad reads the poem, closes the book, and begins talking about all the wonderful novels I am going to write. 'How about one called *Alice Back through the Looking Glass*?' he proposes, then looks up at me, surprised by his choice of title. He says nothing about the poem. Mum reads it carefully several times. 'Such a strange poem, Barb, what is it about?' But I won't say any more. I am sitting at the kitchen table with her and she smiles at me. I know that smile well; my muscles tighten. She reaches out a hand and strokes my cheek. 'Dear, I know that you are mentally unwell. I know that, Barb,' she says, still smiling, 'because you won't explain your poem to me.'

My mother, I told V on our first meeting, is my best friend. She understands me better than anyone else, better probably than I understand myself.

A woman is rushing down a city street, holding a little golden-haired girl by the hand. She pushes the child into my car and hurries on. I drive my car into a nearby street and park and walk away. But what about the little girl? I hurry back to the car but I can't see the child: where is she? I dig down into a pile of garbage beside the car and pull out a golden-haired doll. But no little girl – she has vanished. I walk through the city, searching for the child. The streets are concrete and granite, treeless, empty. I come upon a small amphitheatre where a group of people is standing around. They are members of a cult which kills little girls, cracks open their skulls and eats their brains. I walk among these people and at first they aren't perturbed by my presence. Then they begin to become suspicious. I look around frantically for the little girl but there's no sign of her. 'What are you doing here?' a man asks me. 'I'm looking for my mother,' I tell him. 'Your mother?' he laughs, and I see that my mother is standing next to him, holding a knife to her chest. 'Where's the car?' she shouts at me. I scream back at her, 'Where's the child? You have killed her!' 'There was no child,' she tells me. 'There was no child, there never was a child, there is no child,' the crowd chants at me – and I wake, heart pounding.

I write this dream in my journal, hiding away in my bedroom in Saskatoon. As I do so, I am suddenly overcome with fear that Mum and Dad will read it and discover just how insane and dangerous I am. They will have me locked up; they will refuse to allow me to return to London;

they will keep me for ever in this daughter-hell they have created for me. Maybe I should kill them first? This fantasy scares me so much that I want to run away, and I hurry to reassure myself that these are just thoughts, and that Mum and Dad will never read what I've written. As I'm telling myself this, the door swings open and Mum walks in. I barely have time to hide my journal. 'Hi, Barb, what are you doing?'

When I was thirteen Mum found my diary and read it through. She told me about this a few years later. 'I was sure you wouldn't mind,' she said comfortably. 'It was very interesting.'

For the rest of the visit, I keep my journal in my shoulder bag.

When I was a kid I used to worry all the time about Mum.

What do you mean, worry?

That something would happen to her. I could hardly bear her being out of my sight. Even in a supermarket, if I couldn't see her I would get really frightened. It was terrible, especially at night, if she went out.

How old were you?

Ten, eleven?

As old as that?

Yes, maybe even older. It went on for years, especially the anxiety when she went out at night. Sometimes it made me sick, I would throw up.

Telling V this, I see the scene so vividly that the consulting room fades; I am back in my dark bedroom, lying rigid in bed, waiting, waiting, waiting. The headlights from a passing car move slowly across the wall opposite the window . . . my breath catches . . . I wait . . . The car passes. Another set of lights . . . and another . . . the minutes and hours pass like this . . . Then lights slow . . . and stop. I leap out of bed and race to the window. The car engine is turned off . . . Is it her? I peer through the dark . . . no, not her car . . . I sag back from the window, panting with disappointment and fear. My brain is burning, my stomach churns . . . how will I survive this night?

This nightmare repeats once, maybe twice a week. I never speak about it; I know it will anger Mum.

Now every weekday I wait with pounding heart to see whether V is still alive. One day during a session I become so frantic with fear for him that at the end of fifty minutes he picks up a book from the small table next to him and hands it to me. 'Bring it back tomorrow.' I walk down the street away from his building, hugging the book and crying with relief.

5

Dead Babies

There's something between us that isn't there . . .
A mother describing her abusive relationship with her baby son[1]

For the times, Mum was an elderly mother; she was twenty-seven when I was born. Dad, she told me later, hadn't wanted children when they married, and anyway they couldn't afford them. Their early married years had been tough financially, with her supporting them both by working as a secretary while Dad attended law school. But as soon as Dad was settled into Sede's law firm, Mum got pregnant. She was triumphant. 'Look, Daddy,' she recalls saying to Sede, showing off her big belly.

I was born, so the story went, at one minute past midnight, and emerged 'covered in gold' (I never asked what this meant). My sister, Lise, was born two years later. Soon after Lise's birth Mum suffered what she later described as 'the miseries'. In fact it seems likely that she had a full-blown breakdown. She became, on her own telling, a hazardous mother, tumbling downstairs with me in her arms, putting me in a bath and forgetting about me, feeding Lise a plum which nearly choked her. Later, on summer holidays at the lake, she would send us out to the playground with a warning to watch out for bears (the bears were no fantasy – this was in northern Saskatchewan). 'I don't know what I was thinking of,' she would chuckle in later years when I reminded her about the bears.

As I got older, the danger passed from her to me. Life was impossibly treacherous; any movement could bring catastrophe. If I ran fast or jumped high Mum would scream: 'Careful! You'll hurt yourself!' Once she grabbed a knife from me and almost cut my finger off. Every ache, every cough, saw me in the doctor's office. Death was just a heart skip away.

Every new mother occasionally wishes her baby dead. Today, women who wish it so much that they risk acting on it have a diagnostic label – 'postpartum psychosis' – and a battery of remedies. Things were different back in 1952. When Mum consulted her doctor he told her she was feeling bad about her failure to give birth to Lise 'naturally' (her cervix hadn't dilated sufficiently and drugs had been required). She should just let the guilt go, he instructed her, and focus on the joys of motherhood instead. I'm sure Mum tried – she believed strongly in the joys of motherhood ('Your birth was the happiest day of my life, Barb!') – but it didn't work. So she signed up for law school and handed over Lise and me to a young woman hired for the purpose. I was two and a half, Lise was six months old. For the rest of our childhoods we shared our home with a 'housekeeper'. We took this for granted, as we did Mum's partial withdrawal from our care and Dad's complete abstention from it. It was just how things were, and as I got older it became another sign of our familial superiority. Only dull, ordinary families had stay-at-home mums. (Dads who enjoyed spending time with their kids, as portrayed on the popular TV show *Father Knows Best*, were so implausible that I dismissed them as Hollywood inventions.)

Our first housekeeper was Rose, recruited from a local charity home for unwed mothers. Rose arrived with a big tummy and departed briefly a few months later, returning flat-bellied. (I have a memory, which is probably apocryphal, of seeing her returning from the hospital dragging an empty case.) She remained with us for another two years, before leaving to get married. Encouraged by this success, Mum turned again to the charity home, and a string of pregnant women followed – none of them, in Mum's view, up to Rose's standard. All were country girls sent to the city to avoid scandal. Their babies were invariably given up for adoption. One after another, these girls would arrive, stay for a while, go off for a few days, return – *sans* belly, *sans* baby. Few of them, Mum complained, knew how to cook, and many were 'low-spirited'. Eventually Mum gave up on the maternity home and began hiring through the local press – mostly farm girls, as before, but ones unburdened by illicit pregnancies.

The 'unwed mother' of the 1950s – condemned by moralists, bundled into an institution, robbed of her baby – was a commonplace tragedy. I have friends whose mothers were forced to give them up at birth, and I have met women of Mum's generation who still ache with the excruciating pain of handing over their babies. The cruelty was extraordinary, unconscionable, and went mostly unacknowledged – even by sexual liberals like my parents. It seems never to have occurred to Mum that the low spirits displayed by these girls might be grief. Why should it, when such things were never discussed? And why not give a girl, who has just lost her child, responsibility for your own little ones? She was doing the girls a favour, Mum thought, and no one would have disagreed.

But what happened to the babies? Now here was a puzzle for a little girl. I knew that there *should* be babies because Mum too had grown a big belly which eventually produced Lise . . . But no one said anything about these girls' babies – so what had become of them? Had the babies died, and if so, who had killed them? I don't remember pondering such questions, but after I had been in analysis for a time I started having dream after dream about dead babies, often in the guise of chickens or other small, consumable animals. One morning in a session I suddenly recalled Mum reading 'Hansel and Gretel' to me when I was about five. I loved fairy tales and sat snuggled up next to her, tingling with anticipation. As the story progressed I began to feel anxious, then very frightened. When she reached the bit about the child-eating witch, I howled and struck the book from her hands.

Mum brought home a cat when I was six. Four kittens soon arrived, of whom three were given away. The fourth, which I called Beauty, was killed by a car when it was a few months old. The mother cat then abruptly vanished. Many years later Mum told me she had taken the cat to the vet and had it put down.

> *Mum killed our cat.*
> *What?*
> *We had a cat, she had it killed.*
> *Why?*
> *She said it was dirty.*
> *Dirty?*
> *I think she meant it shat a lot. Mum didn't want more kittens in the house. The cat would have had to have been spayed, and she didn't want to pay for that. So she had it put down.*
> *No more babies?*
> *Yeah. No more babies.*

I talked about the cat early on in my analysis. But it took more than a year for me to remember the housekeepers and their vanished babies. They surfaced slowly, grudgingly, like long-buried corpses rising from a dark pit of fear and guilt. Once resurrected, they haunted my sessions – especially our first housekeeper, Rose, whose reappearance was preceded by days of deadly misery.

> *I feel horrible. I don't know what's the matter with me. I feel really horrible.*
> *Something is bothering you?*
> *Hmm.* [Long pause] *I dreamed I had a baby. A baby girl. I don't remember much about it.*
> *A baby girl?*
> *Yeah.* [Long pause] *If I ever have a baby girl, I will call her Rose.*
> *Oh?*
> *It's my grandmother's name . . . well, sort of. Her name was Rosina.*
>   [Long pause] *I think I was a bit afraid of Grandma . . . In fact,*

I'm not sure I liked her very much. [Long pause] *Rose, Rose . . .*
Rose was also the name of a woman who looked after me.
Oh? When?
When I was very little. Just after Lise was born.
Oh? For how long?
Two or three years, I think. Yes. I think that's right. Mum hired her. I
  don't know what became of her. She was a nice girl, I think. Lovely
  actually. I wonder what became of her. She had a baby.
You mean she brought her baby with her? Really?
No, she had a baby while she was with us. She was pregnant when
  Mum hired her, she was living in a home for unwed mothers. She
  had the baby while she was with us.
Oh, I see. What happened to the baby?
What? The baby? What happened to the baby?
[WHAT HAPPENED TO THE BABY? WHAT HAP-
PENED TO THE BABY?]
Yes, the baby . . .
. . . I don't know. I think it was adopted. Yes, it must have been
  adopted.
[WHAT HAPPENED TO THE BABY?]

I leave this session feeling ghastly. That night I am woken by a cry:
mine?

I dreamed about a wolf last night. Horrible. It had a hat on. Great
  big teeth . . . horrible. And this hat. A green hat? No, pink, a
  pink hat.
Pink for girls?
I guess so . . . a girl wolf? Maybe.
A wolf who eats little girls?
I don't know, maybe . . . Pink . . . what's pink . . . ? Pink for girls,
  pink flowers, pink roses . . .
Ah, pink for Rose, a dream about Rose, and her missing baby?
Hmm . . . maybe. [Long pause] *I feel a bit sick.*

Oh? What kind of sick?
I don't know. I feel funny. Actually, I feel horrible. I feel really
  horrible.
Thinking about Rose and her baby?
I guess . . . I don't know why. She was a nice woman. I'm sure she
  was a nice woman. Why would thinking about her make me feel
  so bad? [Long pause] *I feel really bad . . . horrible. Why do I*
  feel so horrible?
I don't know. We'll have to find out.

The following summer I went again to visit my parents. (This was to be my last visit to them for eight years.) I quizzed them about my babyhood, and about Rose. They were bemused but happy to talk. Rose, Dad said, was 'very striking, a beautiful girl', 'very motherly'. She was a 'big help to me', Mum said, remembering how Rose would take Lise into her bed at night to feed her, keeping the baby with her until she slept. 'Rose was terrific,' Mum said, 'I don't know what I would have done without her.' This thought launches her into another catalogue of her dangerous behaviours with us children ('my stupidities'), including taking us out in a rowboat on a windy day and finding herself stranded in the middle of the lake without enough strength to row us back. It was Rose who took the oars and got us home, she recalls now: 'God knows what would have happened without her!'

Rose left us quite suddenly. She went off to work for another family, and we children didn't see her again. Was I upset? I asked my mother. 'No, I don't think so, well, maybe a little . . . You weren't very demonstrative, you know? Not openly affectionate . . . not that sort of child.'

I have looked for Rose ever since.

When I was nineteen I saw Roman Polanski's film *Rosemary's Baby*, about a young woman who is impregnated by the Devil. Her husband – the deliciously devilish John Cassavetes – has made a pact with a satanist coven to lend them his wife's womb for the gestation of their new messiah in return for help with his flagging theatre career. Mia Farrow played the betrayed Rosemary. I watched in fascination as she was drugged unconscious and raped; as her violated body swelled, and excruciating pains assailed her. She has an emergency caesarean and is told that the baby has died. But Rosemary hears her child's cries and tracks him down to a neighbouring apartment inhabited by the satanists. She reaches the cradle and peers in – and blazing goat eyes stare back at her: her baby, Mephistopheles, evil incarnate. But a mother is a mother: the film ends with Rosemary, garbed in heavenly blue, cradling her terrible progeny.

In fact the film leaves open whether Rosemary is the victim of a diabolic plot or merely crazy. This ambiguity may have been part of the film's appeal to me. In the next decade I saw it three more times, until I knew bits of the script by heart. Then I forgot about it.

The baby is a vampire. She is a zombie-baby, a miscarriage, an aborted foetus that goes on breathing, moving, crying: how can she exist? Yet here she is, a parasite feeding off its mother – voraciously, criminally alive.

I wake with my chest heaving with pain. I drink vodka so that I can get up, begin moving through the day.

Why did she have me? She could have aborted me.
She thought she wanted you.
She shouldn't have had children. Neither of them should; someone
  should have stopped them. I can't bear this any more. I feel like
  I am vomiting myself up.
Or aborting yourself?
Why do you say that? Who said anything about abortion?!
You just did.

6

God

Sex is wonderful. Mum tells me this when I am seven or eight, sitting me down with books and pictures and explaining to me about my vagina and my clitoris. Sex with Dad is fantastic; he is the best lover in the world. This message is implicit, but I understand it perfectly.

Being a woman is wonderful (so long as there's a good man to be had). Mum shows off her breasts to me as she dresses for a party. 'Pretty good, eh? I always had nice breasts.'

Will I be a woman with good breasts? I study myself. I like my skinny nothing-much-at-all body, or rather I am comfortable in it. It tells me nothing about myself that bothers me. I am not ugly or pretty; I have freckles but they don't worry me. Will I ever have boyfriends? I am sure that I won't. As I get older this conviction hardens. We go to London when I am thirteen and I have my first menstrual period while we're there. The pain is terrible, it makes me throw up. I am sure that the blood is leaking everywhere. We're sitting in a restaurant and a boy at the next table glances over at me. Mum sees him looking and chortles, too loudly. 'Look, Barb, he's staring at you. Look, George, that boy is staring at Barb.' The boy hears her and his mouth twists, he turns away. I am sick – with menstrual cramp, but also with a vicious despair that is new to me.

*

I took a small Bible with me on this London trip. I hid it among my clothes, but Mum found it. She left it next to my bed and didn't say anything about it.

Bibles weren't part of our family reading. Like most left-wingers, Mum and Dad were devout secularists. Their bookshelves were crammed with Marx, Lenin, Darwin, little pamphlets from Moscow with titles like 'The Case for Scientific Socialism', 'Historical Materialism and Its Enemies'. As a child Mum had been enthusiastically Jewish, but Dad and Communism changed all of that. Dad himself was a lifelong atheist whose 'conversion' to Judaism after their marriage was merely an exercise in family diplomacy. He never set foot in a synagogue except on family occasions. So I didn't feel myself to be properly Jewish, and anyway I disliked the Jew's outsider status. I preferred to run with the pack. I was uncomfortable when friends referred to my Jewishness, and I hated it when my teacher sent me home with a note for Mum asking whether I should be excused from the daily Bible story. Mum just laughed, what did she care what fairy tales I was told? We had a Christmas tree, although no angels were permitted to perch on it (angels or no angels, Sede wouldn't enter our house when the tree was up).

But we had plenty of religion in our home: Rose, and most of the young women who succeeded her, were Mennonites. And in the year before our London trip, when I was twelve, one of these Mennonite girls set out to convert me.

Mennonites are Anabaptists, followers of the sixteenth-century Dutch Anabaptist Menno Simons. Anabaptism (the label refers to the rejection of infant baptism) was a radical variety of Reformation Protestantism that in its early years was fiercely persecuted by state churches across Europe. In the seventeenth century the Mennonites, like their fellow Anabaptists the Amish, began fleeing Europe for North America. Waves of migration followed, with a peak in the late nineteenth and early twentieth centuries when tens of thousands of Mennonites travelled to western Canada. Most settled in Manitoba, but several thousand went further west to the wheatlands of Saskatchewan. By the time I was born, in 1950, the countryside around Saskatoon was liberally dotted with Mennonite farming communities.

Traditional Mennonism is a religion of prohibition. Believers are forbidden to drink alcohol, dance, watch films or television. Anyone who violates these injunctions is 'shunned', a particularly nasty form of excommunication in which the apostate is forbidden all contact with other Mennonites, even family members. Giving birth to an illegitimate child left a woman shunned for life. The Mennonite girls who came to work for us were from poor homes; they wore cheap cotton dresses and clumpy shoes; their hands were usually chapped. Few had more than elementary education. Mum was contemptuous of them (except Rose), and made little attempt to hide it. Most stayed with us for only a short time. One, a pale, skinny girl, wrote a long letter to Mum after she left, begging forgiveness for stealing a package of chewing gum. Another stayed in her room, weeping. But one, a bubbly girl named Erma, became my chum. Erma had backsliding tendencies; she yearned to wear pretty frocks and to go dancing. Spying an opportunity one day, I persuaded her to take me to an Elvis Presley movie. Even with her recidivist streak, it took all my persuasive powers to get us to the cinema. Once there, as the lights went down, I felt her stiffen beside me, then start to tremble. Appalled, I began trembling in turn and the two of us sat there, quivering with guilt and misery, for the entire film. I don't think she ever forgave me.

Then Anne arrived. Now here was Mennonism with a difference. Anne was in her early twenties, bright-faced and intelligent. She was also, despite her youth, ardently evangelical, a determined fisher of souls. I liked her immediately. I like to think that she also liked me, that something in my young self attracted her apart from my spiritual condition. Perhaps so; but my lonely unhappiness made me a sitting duck for conversion, and Anne's campaign began almost immediately.

Books were her weapon. I would devour anything with a plot. We read the Bible together and chatted about it. Then, having discovered what a story-junkie I was, Anne began feeding me teen novels featuring a 'Spirit-filled' boy named Danny Orlis whose otherwise dull adventures were enlivened by regular encounters with reprobates who, dazzled by his piety, would fall to their knees and 'declare for Christ'. The plots were so thin and repetitive that I was soon bored by them, but Danny's born-again status, and the power it conferred on him, fascinated me. Even adults hardened by years of sinning would fall under Danny's spell. The notion transfixed me, as did the peculiar sexual atmosphere of the books, with their chaste flirtations pepped up by mystic ecstasy. The idea of boys and girls mutually convulsed by divine love seemed odd, but it spoke to something new in me, a disturbing inner pressure that filled my daydreams with strange images and turned my nights hot and gaudy. Were these signs of satanic influence? Was I one of the damned? I started to fear for my soul.

Finally, after some months of chatting about Danny and his adventures, Anne sat me down and asked me the question I'd been expecting. She knew I had been thinking a lot about my Heavenly Father and His great love for me; did I feel ready to accept Him into my life? I hemmed and hawed. I didn't know, I wasn't sure – could a Jew really become a Christian? Of course, she assured me. Well, OK then . . . She hugged me and I wanted to cry. I liked her so much and she cared about me, she wanted to help me. I shouldn't disappoint her, I should do what she asked. But she felt my uncertainty and didn't press me. Maybe I would like to talk it over with God? That appealed to me and I began writing long letters to Him, telling Him about my life and confiding my spiritual doubts. I offered to show the letters to Anne, but she told me they were just between God and me.

Many pubescent children get the God bug – even, or maybe

especially, children from atheist homes. Religion, especially funda-
mentalist religion with its lurid symbolism and moral stridency, is a
better fit with adolescent extremism than the dry-as-dust secularity
in which I was raised. My early teen years were a maelstrom of confusion. Everything, according to my parents, had a rational
explanation; yet our home shuddered with the inexplicable. Why did
Dad yell at Mum all the time? Why didn't Mum yell back? And what
did they do in their bedroom every lunchtime? I never consciously
thought these questions, much less asked them, but by the time Anne
entered my life they were pressing hard on me. I began acquiring tics
and compulsions. My body felt acutely uncomfortable, not anywhere
in particular but everywhere, as if I had grown an all-over hair shirt.
Then in the months after we returned from London I developed a
washing compulsion, which soon became acute. This was strange,
and I knew I had to keep it a secret.

I had never been much of a washer. In my first year at school, when
I was six, the teacher had begun the day by asking who had washed
their face and hands. I had always put my hand up. Why not? I was
sure everyone else was lying too. OK, maybe the hands after using
the toilet, but the face – who would ever bother with that? I never
felt dirty, and I hated wasting time on useless tasks (like dressing:
aged seven, I wore my clothes to bed for several days until Mum,
spotting my shoes sticking up under the covers, pulled them back to
find me fully kitted out for school). But now I began pouring soap
powder into my twice-weekly bath. Every bath time I poured in more,
stirring the water vigorously until the soap grains dissolved. A few
weeks of this, and a fierce rash appeared across my belly. The doctor
stared at it, puzzled – had I been putting something on my skin? I
didn't tell him about the soap powder, but I stopped using it and the
rash faded.

My genitals had always seemed nice and clean – soft little lips, with
an edging of creamy whiteness. But now they became foul and stinky.

46

I couldn't actually smell anything, but I was sure the stench was there. And I couldn't bear the feeling of the vaginal lips touching each other. Sitting, I would twist and turn; walking, I would stretch my right leg to one side until I could feel the lips parting. 'For heaven's sake, Barbara, stop lurching about!' Mum was exasperated. 'What's the matter with you?' Menstruation confirmed my filthiness, and then there were the pimples – disgusting! My skin was so greasy, it was unbearable. I began washing my face with an anti-acne cleanser and soon I was washing it dozens of times a day. In school, I raced to the washroom between classes. My friends began to notice this and to tease me, but my putrid skin could not be tolerated and I ignored them.

Christianity is strong on dirt. Anne took me to her church one Sunday, a little clapboard building on the wrong side of the tracks. The minister, a young man with ferocious acne, stomped up and down, railing against sinners who handled the 'beautiful toys' of life with 'filthy hands'. Did we care how dirty we were? he hollered. Did we want to cleanse our souls from sin? Anne glanced sideways at me. Did she know about my washing? A few young people were going forward, readying themselves to 'declare for Christ', and I saw she was hoping I might follow suit. But dirty as I was, I wasn't ready for that yet. And soon Anne was gone.

One night I stayed over at a girlfriend's house and spent the evening regaling her with tales of sin and damnation. I didn't like her parents, who were bohemian snobs, and I luridly described the hellfire that awaited them. The next morning the poor child, hollow-eyed, reported the conversation to her mother. Mum was rung, and a few days later Anne told me that she was leaving us, going off to a training institute for missionaries.

So it was my silly preaching that lost me Anne; I felt sure of this at the time. But there is another memory that tugs at me. Perhaps it was just a dream. I walk into the kitchen to find Dad there with Anne. Anne's face is flushed and strange. Dad looks around, sees me and walks out. Anne says nothing; I ask nothing. A few days later she tells me she's going to leave. I weep a bit, and she holds me. 'I don't want you to go. I don't want you to go.' She hugs me hard. Then she says, 'Your father . . .' and stops.

My Dad was a philanderer, a grabber. He had many lovers. 'He was quite something,' an old friend of Mum said to me a few years ago, as she described Dad arriving at parties in Saskatoon, accompanied by Mum and his mistress of the moment. 'He just flaunted them . . . I don't know how Tillie could stand it.' As a child and adolescent I refused to know about any of this, eyes wide shut, even when Dad paraded his conquests before us; even when, in my university years, a friend at a gathering where Dad was present suddenly hissed at me, 'Your father needs to keep his hands to himself!' Still I heard and knew nothing. The knowledge pulsed far down inside me, along with my jealous fury at Dad's betrayals: a purulent secret so deeply hidden that it took almost a decade of psychoanalysis to disinter it, its presence marked only by its filthy seepage into my dreams and on to my disgustingly filthy face.

What I *did* know about, what I permitted myself to recognize, as I travelled through puberty, was the change in Dad's demeanour towards me. Up to then he had been mostly distant or irritable, with occasional flashes of warmth. Now he was angry with me most of the time. Everything about me seemed to grate on him: my behaviour, my appearance, my opinions. He became absurdly dictatorial, barking orders and shouting if I didn't jump at his command. 'Just like with Mum,' I thought, with a touch of pride. He was also over-intimate, touching and stroking me whenever I was within arm's length. All this left me disturbed and vaguely frightened, but excited and prickly too. Our dinner table became a war zone, with Mum and Lise cringing on the sidelines.

Anne left, and I began high school. Eighteen months later I fell into an emotional pit. A nauseating boredom engulfed me, along with an intense loneliness. My fights with Dad grew savage, until one evening he ordered me out of the house. I wandered the streets for an hour or so and then sneaked back in. I tried to talk to Mum about my misery but she was impatient. 'Kids always exaggerate,' she told me, and I knew that it was true, that I was absurd and histrionic. I loathed myself and I wanted to die.

In the midst of this unhappiness I heard from Anne. She wrote to me from her missionary institute, letters full of tender queries about my schoolwork, my holiday plans, my bible-reading. She worried that I was missing her, and reminded me that she and God were always glad to hear from me. 'God wants us to talk to Him a lot. That's why God made man in the first place, to have someone to love and talk to. But man sinned and spoiled it all, so He had to send Jesus. Wasn't that a wonderful love?'

I recalled Anne about a year into my psychoanalysis, when I experienced what she would probably have described as a visitation. I was sitting in my study on a winter afternoon. My housemates were all out; I was alone in the house. Then I was not alone. Someone was in the room with me; I was surrounded by a presence that enfolded and embraced me. The feeling grew, and with it a certainty of my invulnerability: I was blessed, nothing would ever go wrong in my life again. The feeling came and went for several days, and then disappeared entirely.

What do people want God for, what use is He to us? 'There must be a God,' Martin Luther is claimed to have said, 'because man needs one being whom he can trust.'[1] I never prayed to God when I was ill, nor did I pray to V, although my yearning to place my trust in him, to feel myself unequivocally loved and cared for by him, was immeasurable. But the unbeliever in me knew that such care was impossible – as impossible as God's love. Perfect love and perfect care were impossible, which meant that love and care in general were impossible – for how could there be imperfect love and care? (The discovery that love and care are always imperfect, when I finally got to it, was an extraordinary liberation.) And love and care must be impossible, or why hadn't my perfect parents loved and cared for me? 'People use God like an analyst – someone to be there,' a young woman patient once said to the British psychoanalyst Donald Winnicott. '[Someone] for whom you matter,' Winnicott replied. 'I couldn't say that . . .' the woman replied, 'because I couldn't be sure . . .'[2]

7

Sex

Sex was a filial duty. Fucking men was what my mother's daughter did. I didn't have boyfriends when I was in high school but I knew this had to change when I got to university. When I was seventeen Mum took me to the doctor for contraceptive pills, and at eighteen I went to Montreal to lose my virginity. There I met a man who bedded me in a converted broom cupboard. He was thick and hairy and his genitals repelled me, but I behaved as I thought a sexy woman would. This man didn't approve of sex so my clumsy fakery suited him. My next lover liked sex, and occasionally brought me to orgasm. But the whole business was messy and not much fun – and anyway, what was the point? There was only one man worth having, and that was Dad. To imagine otherwise was impossible. My lovers, Mum tells me, are 'immature'. I do not disagree; Dad is The One, he brooks no rivals.

So my lovers were required to disappoint. But I had to have them, single aloneness was unthinkable. In 1971 I arrived in London in a flat panic – what am I doing here, in this terrifying city? How am I going to survive? – and grabbed the first man in sight. We had nothing in common except self-dislike, but I stuck with him for a year or so. Others followed, but nobody to beat Dad, nobody who even came close to him. I knew where my loyalties lay. A couple of years later Lise married a man she loved deeply (and still does); Mum and Dad denigrated him constantly.

I never lived alone with a lover and I never contemplated marrying one. I hated the idea of marriage and tried to persuade Lise against it. Partly this was the mood of the times. Women's Liberation was running strong and none of my friends, all active feminists, would have dreamed of marrying. Most of us spent the 1970s in houses run as left-wing collectives. My first home in London was a little flat, but within two years I was living in a big squatted house in Brixton with my partner and a shifting population of students, film-makers, and full-time political agitators. A few years later, after a series of unsuccessful battles with the local council over squatters' rights, some of us moved from this house into another squat, a tiny place in Clapham with an outdoor toilet and a bath in the kitchen. My partner and I and another man and woman all packed into this house; it was too tight and tempers frayed. Finally the other woman bought a bigger house across Clapham Common and we abandoned squatting for owner-occupation. By now everyone was in their thirties with jobs, and the bourgeois lifestyle we had once scorned (or in my case, pretended to scorn) was looking increasingly attractive. So in 1982, shortly after I began analysis, I moved with my partner and another couple into a big, handsome house in Highbury: so much for left-wing austerity.

My partner was intelligent and energetic, and for nearly a decade we got on well enough. But he was never going to be the Real Deal. Mum had bagged the only truly desirable man and I was stuck, as the rules required, with an after-ran.

However, with mental breakdown and the advent of V, matters took on a different hue. Now V was my not-Dad man, the childish dunce whose asininity served as a foil to Dad's virtues. With V as my clay-footed idol, I was free to seek lesser heroes. And I did, leaving my partner and eyeing up other men in my circle, bedding many. Most were left-wing academics, writers, journalists; nearly all were married or otherwise attached. For three feverish years, I dashed from one affair to another, dreaming about wife-murder, mother-murder. I made women furious with me, including some of my friends who found my behaviour astonishing. I was as surprised as they; I had no idea where this Jezebel had come from. Two of my lovers had the same first name as my analyst. 'Common name,' V said.

I am desperately infatuated with Jack, my Irish-American lover. Like several of my previous lovers, he superficially resembles Dad, being a working-class lefty of strong insecurity and inflammable temper. Humming with incipient violence, Jack makes me feel sexy and dangerous. He has a wife from whom he is semi-detached, and a baby daughter, whom he adores. Our affair ignites with a bang – he is compelling in bed and a great storyteller, captivating me with tales of lowlife America – before rapidly descending into a nightmare of frustration, rage, mortification. Jack woos me passionately, and then withdraws; he rings me from his wife's house to cancel our dates. One morning, after a long and lovely night, he announces that sex with me is over; our affair must become platonic. He cannot simultaneously be my lover and a good father to his daughter, he explains with sombre matter-of-factness, looking pleased with this upsurge of paternal conscience (which soon subsides). As I weep with bewilderment, he glances over at me slyly. 'You are too much for me, Barbara.' He is certainly too much for me. Pain blossoms inside me. I am overwhelmed with furious despair, and although I gradually perceive the sadism behind his behaviour, I cannot let go of him. I cry, I telephone, I strategize desperately. My pill and booze intake rise sharply. I feel like I'm going to crack. Yet inside all this mayhem, I am deeply puzzled. Why him, why this man? In my analytic sessions I rail against Jack – against his cruelty, his macho posturing, his crass sentimentalism.

Do you like him?
[Do I *like* him? What kind of question is that?] *I'm in love with him.*

Is 'love' the word for this torment?

We are lying under an open window on a hot summer night. Light from a garden lamp carves our flesh into blue curves. My head is against Jack's neck; I listen to his breath. His hand moves slowly across my heart and downward, taking my heartbeat with it. I feel my skin pulse heavily against his fingers. He sighs. I pull away slightly so that I can see his face. 'You O K?' His eyes glint; he raises himself on his elbow and looks at me.

'O K, Barbara, now we're in LA.' 'Oh yes?' 'We're going to get up, and walk over to Atlantic Boulevard. They have the best tacos there, the best in California.' 'At this hour?' 'Are you kidding? This is LA. At any hour.' His hands move through the air, mapping LA for me, our new lives there. I lie beside him seeing it all, so ready for this, this California dreaming.

We are at London Zoo in Regent's Park with his daughter. Annie is tiny, a gorgeous black-haired chick. Jack hunches over her like an anxious hen. He doesn't look at me, keeps walking away from me. When I offer to hold Annie, he lurches back in panic. He has suggested this outing, but he can hardly stand it. I am so angry and miserable that I am shaking; I need a drink. Annie however seems happy enough, and I feel myself hating her – her wide, remote eyes and clinging hands. I want to drop her in a nest somewhere and take Jack back to bed.

Suddenly Jack begins talking about his wife, what a wonderful mother she is, how much Annie loves her. He goes on and on; then turns to me and leers sentimentally. 'I'd love to see you pregnant, Barbara. Do you think you could hack it?' His voice is so sugary and aggressive that I nearly break down. What on earth am I doing with this man?

Hating him. That's what you are doing.
Why?
That's what we need to find out.

*

Mum visited me in London when I was twenty-two, and took me to lunch at Harrods. I had been in the city for six months, and was proud of my metropolitan nous. I had a flat, I could find my way around on the Tube, and I had two lovers – the height of sophistication. I had written to Mum about the lovers, as I wrote to her about everything. Now, over our avocado salads, Mum told me about her lovers. 'Your dad had girlfriends, so I thought, why shouldn't I have boyfriends?' She wouldn't tell me the names of these men, or of Dad's girlfriends, and I didn't try to guess them. I can't remember much of the conversation after that; I think my head floated away at some point. I know I was struck by how much Mum enjoyed telling me these home truths: she gave me a big hug afterwards. 'It's so lovely to talk to you like this, Barb. You aren't always easy to talk to.' But she was wrong; I was much too easy for her to talk to.

Mum's boyfriends, I later concluded, were fewer than she implied – she didn't want me to think I was getting ahead of her – and acquired mostly in revenge for Dad's affairs. In the late 1960s, a few years before I left home, Dad settled into a steady relationship with a young lawyer in his firm. Gail was clever and dainty, and seemed never to put a foot wrong with Dad, who was often very irascible with younger colleagues. She became a family fixture, eating most of her weekday dinners with us and spending weekends with Dad on out-of-town work trips. I told myself that she was just Dad's colleague, and went on believing this long after Mum's Harrods revelations.

In the summer of 1972 I visited Saskatoon and found a young man named Simon half living in our home. Simon was a handsome Australian whom Mum had jailed briefly for some minor offence a couple of years earlier. He had sent Mum flowers from prison, with a note thanking her for 'turning him around'. She had invited him to visit her on his release. Now here Simon was, seated at the dinner table alongside Gail. I must have known what was going on but it was years before I admitted it to myself. 'Oh, so you guessed?' Mum said when I referred to 'your boyfriend Simon' about a decade before she died.

She smiled, still pleased with herself. When I mentioned the exchange to Lise she told me about an earlier conversation between her and Mum in which Mum described fucking Simon in front of Dad. 'She said Dad liked that.' Mum had had a stroke a couple of years before. She had made a good recovery but still got muddled at times. We decided she had probably made up this story about Simon and Dad.

My parents, I tell V in an early session, are devoted to each other, which in a way is true, although the sadomasochistic elements of their romance remain buried inside me for many years. 'The children of lovers are orphans,' Robert Louis Stevenson says,[1] and this is surely even truer when love is wedded to cruelty. The pain that I felt as the truth emerged was terrible. It took eight years of psychoanalysis before I could remember that Dad had flirted with waitresses – a memory that surfaced on a tide of misery and fear so intense that it almost pushed me out of the analysis. By contrast, I remembered well how Dad had liked to fondle me, pulling me down beside him on the sofa to caress me from neck to knee. He did this until I left home at twenty-one. But this memory, even as I recount it to V, seems ordinary, inconsequential.

Finally Jack and I part, after a scene so humiliating that even I, with my new-found lust for erotic misery, cannot stomach it. I am dressing after an evening of dreary sex. Jack is sitting on the bed, watching me pull on my underpants. He says, 'I want a girl, Barbara. An eighteen-year-old with perky tits, not a fucked-up thirty-something.' His voice is reedy with malice, his face is a study in malignant self-disgust. I bang my way out of his flat, and that is it for me and Jack. The following weekend I go to a party and leave with a skinny artist who beds me with palpable distaste. I have been losing weight and our hip bones grate together. This man is such a flagrant woman-hater that when I boast about him to a friend she is appalled. Yet I continue to see him for several weeks, and when he stops ringing I am overwhelmed by feelings of abandonment. My world is a blank desert; I cannot survive in it. I develop a new symptom: a rhythmic sucking in my mouth, which goes on day and night and makes my throat ache. Writing in my diary, I find it difficult to form the letters, they come out looking like a child's shaky handwriting.

> This pain stifles me. Can hardly write. Encased by this
> aloneness. Can do nothing. Words move nothing. How did
> this pain begin? The slant of light on brick. Cold sun. How
> did this despair begin? Crying alone. Paying a man to hear my
> despair. I am suffocating.

Over the following months I twist and turn inside my loneliness. I drink heavily, take fistfuls of pills, rail at V, whose stolidity in the face of my suffering makes me nauseous with rage. I have more sexual encounters, all of which travel the same emotional path from dizzy excitement and triumph (Another one! Look at me, Mum!) to a sick misery that, even without V pointing it out to me, I now realize has little to do with the men involved. Almost any man, it seems, will produce this suffering, so long as he eventually rejects me. Even men I have slept with only once, even men who bore or repel me, leave me in the same wretched state. My nights fill with dreams of sexual violence. My analytic sessions become frantic as I try desperately to comprehend what is happening to me. In a way of course it is all embarrassingly obvious. I tell my friend Max about V's comment that I seem to be finding versions of Dad everywhere. Max, a sexual cynic who has been watching my antics with distaste, hoots with laughter. 'And for this you're paying him? Pay me – or just read the bloody books.'

Max is right of course – I'm an educated woman; I know all about Oedipus, and the psychoanalytic transference. I can see that it is no accident that my first affair began within days of my first analytic session; and I can easily perceive how Dad and V – such very different men – have merged into my romantic ideal, which I pursue fruitlessly. Measured by the man I want – the left-wing hero, the all-knowing doctor – all my lovers appear seriously deficient, which doesn't prevent me from agonizing over them.

So I am not surprised, I tell myself, by what I am going through (not being surprised is very important to me), although I am disturbed by the intensity of it. What does surprise and upset me very much, however, is the ferocity of my hatred for V, alongside the love and yearning. Ambivalence is unknown to me: my inner world is black and white. I follow strict emotional rules, and hate is not permissible. But hate, I am now discovering, is the breath and bone of me. The exhaustion that crushes me; the aching tension locking my muscles; the tightness in my throat and the weight in my chest: my hatreds are turning me into stone. I dream of slashed bellies, decapitations, mutilated babies. Lying on the couch, I quote a line from a case history of Donald Winnicott: 'It is best in this stage of the analysis if the analysand does not carry a gun.'

8

Hunger

Two years into the analysis V tells me that I am very ill. I am at high risk of suicide, he says. It's a bright springtime morning; some pigeons are flirting on the window ledge outside the consulting room.

What do you mean? What on earth are you saying to me? What
  are you saying to me?
I had to tell you. I didn't know how to tell you.
Very ill! What's 'very ill'? What do you mean, 'very ill'? Suicide!
  Suicide!! You frighten me.
I'm sorry.
Very ill! What are you saying . . .? What's wrong with me?!
I think you have a feeding problem.
A 'feeding problem'!? All this terrible misery and pain . . . a 'feed-
  ing problem'! I'm going to kill myself because of a 'feeding
  problem'? You're fucking crazy; how can you say such stupid
  things to me? You make me sick, you asshole.

My greed was a family legend. I was as skinny as a rake, but I craved and ate and craved more. I was never satisfied. Mum called me a *khazer*, Yiddish for 'greedy pig'. She was stern about candy; my ration was one candy per day. A bowl of mints sat on the coffee table. One afternoon, when I was seven or eight, I took a handful of these mints and put them into my shirt pocket. Mum spotted the bulge – what did I have in there? My breath stopped. 'Bread?' I gasped. She sent me to my room, but the episode disturbed her. In the evening, just before bedtime, she took me on her knee. Didn't I have enough nice things to eat? Did I want her to make more pies, bake more cakes? She knew that other kids, munching treats made by their stay-at-home mums, occasionally taunted me about my out-at-work mother. I was flummoxed and said nothing. My candy ration remained the same.

Later, when I had pocket money, I bought huge amounts of candy. I would stop at a shop after school to load up. At home I would curl up in a chair with a book and my booty. I felt disgusting. One afternoon I ate so many candy bars that I nearly threw up. I stared at the pile of empty wrappers, woozy and vaguely frightened. If this was making me feel so bad, why did I do it?

Nothing you say makes any difference to me!
Yet you come here every day.
Yes — God knows why! I talk and talk, and you talk and talk, blah blah blah, and nothing makes any difference!
What kind of difference do you want?
I want to feel better! What do you think, you bastard? I feel awful, I feel sick.
I think you want to feel fed. But you are afraid.
Fed?
Yes, fed. Full, satisfied.
[Long silence] Why afraid?
Because if I ever find out that you feel fed by me, I will make you suffer for it. That's what you believe.
Suffer? How? What do you mean?
Suppose you left a session feeling really satisfied by the work we had done. What do you think would happen next?
What? What are you talking about?
You believe that I would stop doing whatever I had done that satisfied you. That's what you think would happen. And then what?

Stop doing what? What would you stop doing? You mean stop seeing me? Stop analysing me?

Stop analysing you, stop feeding you, stop seeing you . . . yes. That would be your punishment for feeling well fed by me.

That would kill me.

Yes, you would starve to death. That is what you believe.

Hunger consumes me: not physical appetite, but a craving for something to be put into me that will still my anguished voracity. This gratifying thing, whatever it is, is the Real Thing that other people have and I do not. Knowing that others have it drains all pleasure from my life. My envious rage consumes me. I see them there, looking so content, so pleased with themselves, and I feel nothing but hate. Satisfaction always belongs to someone else – my mother, V, the men who fuck and leave me.

I am starving. Nothing I put into me satisfies me. I could devour the world, but I chew myself (my nails, my hair) instead. When I eat, I keep on eating. There is no repletion, no stopping point. Candy bars, vodka, pills: I keep going until I throw up.

Fucking, and psychoanalytic interpretations, open up the same abyss of craving and rage. No one must touch me: starvation and death are inside me.

'He says I'm suicidal,' I tell Cora. 'Do you think I'm suicidal?' Cora doesn't look as surprised as I think she should. 'I don't think you want to kill yourself,' she says after a moment. 'But I think you want to be dead.'

Now on most weekdays I stay in bed until it is time to go to V. Sometimes I lie there drinking vodka then drive to my session drunk. I leave the car at Regent's Park and walk over to Harley Street. I am always on time, often very early. The building, a four-storey terrace built to house well-to-do Georgians and now given over to medical consulting rooms, is large, opulent. I wait in the reception area, sunk into a padded leather chair, a book on my knee. Glossy Middle Eastern women come and go, fur coats slung over arms, perfumes freighting the air. The receptionist eventually summons me. 'Dr Taylor.' (V always insisted on using my title.) Eyes turn to me, slumped in my chair in my jeans and sweater. I trudge up the stairs to the consulting-room door, knock once. 'Enter.' V is standing beside his chair, behind the couch. He waits while I lie down, then he sits and we begin.

In the course of my analysis I repeated this routine, with minor variations, some 4,000 times. To outsiders, the rites of psychoanalysis can look ridiculous. The hushed room with the ruminative analyst; the padded couch; the fifty-minute hour – popular culture is crammed with satirical representations of these. I had them all. At first I cherished them as cultic icons, symbols of the higher mysteries to which I was an eager initiate. Later I savaged them as cheap tricks, the flimflammery of a charlatan. And later still, much later, I came to see them as containers for the uncontainable, solid supports for emotional chaos. V ran a tight ship. No casual running over time, no out-of-session chit-chats, no personal information about him (except his home phone number, which I mightily abused). Over the years I came to lean on this formalism like a rock. I could scream, threaten, arrive drunk and abusive – nothing made any difference. Whatever went on in the session, however desperate and wild I became, at the end of fifty minutes V would call 'time' and I would rise and go. I never tried to stay on – although for a few days I couldn't operate the door handle, struggling with it for several moments after each session before it finally turned and I fled.

V never described me as 'regressed', perhaps because I had never really grown up. But now, in the wake of my affair with Jack, I find myself sliding into a place that is simultaneously deeply familiar and utterly alien. A nightmare landscape opens up inside me, populated by violent cartoon images of my parents, sister, lovers, friends. In my sessions my speech careens between registers, words shattering into nonsense syllables. Rage burns through me. I crash my car twice in a fortnight, which cheers me up temporarily. I chuck down pills like candy. 'Poisoning the little girl,' V says, which terrifies me. I bellow at him to give me something to relieve the fear and pain.

Let me have it! I can't go on like this.
What? Let you have what?
Give it to me! Give it to me!
Give you what? What should I give to you?

The answer screams up through my body, through my pelvis, my chest, my throat. It is a warhead inside me; every molecule in my body is stone-rigid, braced against it.

I don't know. I don't know. I don't know.
Time.

In the autumn of 1984 I resigned my college job, knowing that I would not return to it when my research grant ended. I told people I intended to devote myself to full-time writing. I thought I would probably never work again.

A friend organized an unpaid research fellowship for me at a London university. I was given a small office where I was meant to work on my Wollstonecraft book. I went to the office most days after analysis. Occasionally I read and wrote a bit, but mostly I just sat for long hours, doing little, suffocated by loneliness. I slept with another researcher, a good-looking man with a wife and two small children. Matthew was soft-spoken and depressed, and cared as little for me as I for him. We could hardly look each other in the eye. Nonetheless I was fiercely jealous of his wife, and to spite Matthew for staying with her I slept with another man, then a third. I stopped eating and became very thin.

What the fuck am I doing?
I don't know. Trying to make me jealous?
Hah! [Is he? He looked at me oddly when I came in . . .] *I don't*
   think I even like *fucking men!*

No?

Maybe I'm a lesbian. [When I was sixteen I had had a passionate
  friendship with a girl who turned out to be gay.]

Oh? You feel attracted to women?

*Well . . . no . . . oh I don't know! I don't think I want sex with
  anyone! Not Matthew, not anyone . . . So why I am doing this,
  fucking all these men?*

Maybe you're trying to fill yourself up?

I dream I am feeding a baby boy who does not look at me but instead
looks longingly towards his mother. 'Look at the person feeding you!'
his mother yells at him. I dream of V inserting forks into my vagina.
My aloneness, which cannot grow any more, grows more – until one
morning when, starting down the stairs outside my bedroom door, I
reach out for the banister . . . and encounter a void. No stairwell, no
banister, no one to grip the banister. There is nobody here. There is
panic, but no one who feels it. Nothing lives here. This is not solitude;
it is nullity. I am zero. I sway into oblivion, and then stagger back to
my bed and get blind drunk.

There was nobody there. Nobody! [I am still drunk, and shrieking
  at V.]

Hmm.

*There is nobody anywhere, anywhere! There is nothing but pain. I
  can't bear it, I can't stand it. What are we going to do? What
  are you going to do?*

What do you want me to do?

*What? What do you mean? Did you hear what I said? Are you
  listening to me? Are you listening to me?*

I said what do you want me to do.

*I can't bear it any more! Did you hear what I said? Are you listening
  to me?*

[Silence]

Are you listening to me? Have you gone to sleep? Are you there?

[Silence]

Are you there? Where are you?

You are blotting me out.

What? What?

You are blotting me out.

No, I'm not! What are you talking about?

You can't hear me.

Yes, I can? What are you talking about? I hear you! I hear you!

Do you?

[Do I? Do I hear him?]

Did your mother hear you?

My mother! Always my bloody damned fucking mother! Are you
  listening to me?

I hear you. Could your mother hear you?

[Could she? Could she hear me? Did my mother hear me?]

Oh, oh . . . I don't know. I don't know! How should I know?

Sometime in the week following this session I recalled a song Mum used to sing. She never sang in public situations ('I have such a horrible voice!') but she often crooned to herself while doing things around the house. The song she sang was an African-American spiritual.

Sometimes I feel like a motherless child
A long way from home
A long way from home
Sometimes I feel like I'm almost gone
A long way from home

In the autumn of 2010, a year before she died, Mum spoke to me in a way she never had before. She had been sleeping, and she woke up confused and anxious. I sat beside her without speaking and gradually she calmed; her mood became sad and ruminative. She sighed a bit, and then she looked over at me and hit herself lightly on the chest. 'There is nothing left alive in me that matters,' she said. She had not spoken coherently for some days, so I was startled. Then she said, 'I am afraid.' 'What are you afraid of, Mum?' She looked away and didn't reply. I could feel her gathering her strength. She turned towards me again. 'I have always been alone,' she said. 'I have always felt alone.' 'I know, Mum.' My response seemed to comfort her slightly. Then, with much effort, she pulled herself higher on the pillows, waving away my attempts to help her. She lay for a moment, breathing heavily, then she put her hand on my arm. 'I have always wanted to be a proper person.' She repeated this several times. I touched her hand. 'I know, Mum. I know. And you are – you're a proper person.' She looked sharply at me, as if I might be about to laugh, then seeing me smiling at her she smiled back at me, radiantly.

9

Filth

This should be nothing because it resembled nothing, just as one
minus one resembles zero.

Colm Tóibín, *The Empty Family*[1]

My solitude is bottomless. The further I go into myself, inside this
torture regimen called psychoanalysis, the deeper my aloneness
becomes. It is a fathomless sea with no shore; nothing beyond it,
nothing around it. Things wash about in it. These are my feelings.
Nothing encloses them. A filthy scum lies on the surface. It's a putrid,
sickening scum, but it is the only thing between the aloneness and the
void; it is my aliveness; it is the line between me and point zero.

It's a cold morning. Walking down Harley Street to V's room, I
hunch down into my wool jacket. Some people are coming in my
direction. The wind picks at my neck, lifts my hair. I shudder and
veer towards the railings, away from the other walkers. I should not
be out here on the street, in this bright winter light. I flatten my hair,
pull my scarf closer. Some stray hairs touch my face and I stiffen with
revulsion. I want to tear the hair from my head; slap away the insinu-
ating air. Nothing must touch me. My face is a mask of dungy grease;
filth oozes from my skin. I am an excrescence; I am garbage. As I
approach V's building I feel the pressure welling up: get rid of it!
Wipe it out! A woman coming towards me gives me a quick look.

She is about my age, elegant in a pale beige coat. Don't look at me! Wipe me out!

The hallway off the reception area has a small lavatory to one side. I hurry into it. I pee and wipe, and wipe some more, and then I run the water until it's hot and scrub and scrub at my face. My face is red and raw; my heart pumps thickly, labouring through the crud. I can barely drag myself up the stairs to V's room. There he is, crisply clean, waiting for me – his filthy patient, this lump of garbage.

Before I began psychoanalysis, I never asked myself why I washed my face constantly. It was an irritant and I hid it as best I could, but it wasn't significant; it was just how I was. Back then, obsessive compulsive disorder was a topic for psychiatric textbooks, not reality TV shows and celebrity memoirs. I knew that Lady Macbeth was a hand-scrubber, and occasionally I wondered whether I too had a guilty secret stowed somewhere, but these thoughts never took hold.

By my twenties the compulsion was fixed and I took it for granted. Sex was tricky – all that touching and sweating and flowing; but I washed and washed and took a towel to bed to mop up the fluids. 'You make me feel like a slob, slobbering all over you,' a boyfriend said, as I wiped myself off. But I was the filthy slob whom nothing would purify.

I told V about the washing in my first session and then didn't refer to it again for years. Whatever he might say about it I didn't want to hear. But now, as my wretchedness ratchets up, I need to understand more; and what have I got to lose? 'I need more courage,' I say to V one day. I would repeat this often over the years. In fact usually by the time I got around to asking him a question I already knew the answer (although I still resented him saying it out loud).

Why am I so dirty?
Are you dirty?
I mean, why do I feel so filthy?

You have filthy thoughts. Thoughts that seem filthy to you.

What sorts of thoughts?

We don't know that yet. The sort that you feel on your face. On your dirty face.

My dirty face . . . why my face?

I don't know. Dirty looks?

Whose looks? My looks?

Maybe. Your mother's looks? The way your mother looked at you, when you were a baby?

My mother . . . ? Are you asking me? I don't know! Why are you asking me??

I'm not asking you, I'm making a suggestion. I'm suggesting that your mother might have given you dirty looks; looks that made you feel dirty. Not at the time – but later.

My mother . . . when?

When you were tiny, when she was feeding you. She found it difficult, feeding you . . . it upset her. She gave you dirty looks.

How do you know such things?? You don't know these things!

Of course I don't know *these things!* [Deep sigh] *I'm suggesting this as a possibility: that she was disturbed and distressed, and that you felt this, and it disturbed you very much.*

And made me feel guilty?

Bad, guilty, dirty . . . Feelings that surfaced when you reached puberty, with all its dirty feelings.

Hmm . . . So when I get better I'll stop washing?

Ah. Possibly.

So when am I going to get better?? I can't stand this any more! We keep talking and talking . . . How long has it been now? Bloody years! It's incredible. I feel worse than when I first came here . . . it's incredible.

[Silence]

It's incredible, unbelievable – I keep coming here, and nothing gets better! I just feel worse, I feel terrible! It's unbelievable.

[Silence]
Say something!
[Silence]

I whirl around on the pillow and stare at him. He looks back at me sombrely. I turn away again, embarrassed.

Dirty looks, dirty looks . . . dirty face. I want to feel clean.
Yes, I know.
It's very hard, you know, all this . . . [Weeping]
Yes, I know.

As I leave, he smiles at me. Surprised, I smile back.

I look like my mother. The similarity is striking; people who knew Mum often comment on it. When I was young I liked this but by my twenties I did not, and after I became ill I found any reference to our likeness very upsetting. I am fair; Mum was dark; I am several inches taller than she was. I would focus on these differences, but a quick look in the mirror always defeated me.

I used to sound like Mum as well. She had a loud, cordial voice with a derisory undernote. The derision was mostly directed at herself, an insistent flow of self-ridicule. But it leached into her conversation with others as well, a little drip-drip of mockery – too slight to be openly offensive but leaving its targets feeling vaguely humiliated. Mum found it hard to tolerate signs of well-being. Self-content was alien to her and evidence of it in others could rouse her to real nastiness. 'Look at her, so pleased with herself! [*A little sneering laugh*] That fancy dress – I bet she thinks she's something special!' Feeling special, taking pleasure in oneself, these needed slapping down – in others, but also in her, and in me, her alter ego. To humiliate, to spoil . . . why should anyone feel entitled to happiness when Mum did not?

It's a March afternoon and I am at Sally's house. I'm in her sitting room, supposedly reading, while she works in her study upstairs. But I can't read. I prowl restlessly around the room. Finally I fetch up in front of the mantelpiece, which is covered with photographs, cards, invitations – all the ephemera of a busy, interesting life. There's a mirror and I stand and stare into it. Suddenly my gorge rises, my heart begins to thud heavily. Pain sweeps over me and I collapse on to the sofa. Lying there, my heartbeat drumming me into the cushions, an image begins to take shape in front of me. It is slow to appear, dragging itself out of me: a woman – middle-aged, skinny, sharp-featured. As the woman's face comes into focus, I see that she is laughing. She is laughing at me, pointing at me, hooting at me derisively. As I gawp at her, I realize that I know her well, this bony, malevolent figure – but who is she? The image evaporates and I lie there, my heart bouncing in my throat, certain that something important has happened.

The next day I report the visitation to V.

An image of your mother?
No, no, no – she didn't look anything like my mother. Or sound
anything like her. No, no, no – nothing at all like her!
Perhaps she is a version of you, laughing at me? [This last a touch
bitterly.]

I wave this away contemptuously.

But she is, I soon discover, very much me. The Laughing Woman, I call her, this ruthless harridan who wells up inside me whenever my suffering is acute, poking me, prodding me, delighting in my pain. She is a monstrous edition of my mother; she is my mother's vicious daughter; she is my terrible secret self. Fairy stories and folk tales had already shown her to me in various guises: the witch, the evil fairy, the demon sorceress . . . but above all the hag Envy, with her reptilian tresses and sagging tits, her bloodied mouth. Her presence inside me is vivid, immediate, inescapable. I feel literally possessed, and it terrifies me.

A few weeks later, en route to my session, I spot V walking towards his building. He is smiling slightly, swinging his shoulder bag as he walks. 'He looks happy,' I think. Within moments I begin to feel sick, and as soon as I lie down on the couch I go on the attack. His words, his voice, his appearance; everything about him is wrong and silly; he is utterly contemptible.

What has happened to you??

Happened? Nothing's happened! What are you talking about?

You are a positive whirlwind of destructiveness today. What has happened to you?

Destructiveness? What are you talking about?

Everything about me is bad – the way I look, the way I sound, the things I say. Can you hear yourself?

Hmm . . . yes, well, OK, I hear that . . .

Yes? You hear it?

Yes, well . . . I feel sick today.

I make you sick?

Don't be silly!

I'm not being silly! Something about me has made you feel sick today.

There's nothing about you! You seem fine to me!

I seem fine?

Yes . . . I saw you earlier, walking towards the building. You looked fine.

Oh? You saw me, coming here?

Yes. You looked fine. You looked happy, I thought. I thought you looked happy, pleased with yourself.

Ah.

What do you mean, 'ah'? What does that mean?

What do you think it means?

[I twist and turn on the couch. I cannot bear this. I cannot bear to hear this.] *I hate thinking that you are happy? I hate you being happy?*

'Pleased with myself', I think you said?
Yes, that too . . .
That's what has made you feel sick . . . my feeling pleased with
*   myself, as you imagine it. It's made you sick.*
Sick?
Sick with envy. With envious hatred. You want to destroy me.

A wave of shame cascades over me. I go home, get drunk, telephone
V later to shriek my misery.

I'm dying here! I feel like I'm dying!
You're showing me what happens when I make a correct
*   interpretation.*
What? What? What interpretation?
Today. The reason why you felt sick in today's session.
I don't remember. I don't know what you're talking about.
No. You don't remember because you have destroyed it.
Destroyed what?
My interpretation of your sick envy, of its effect on you. You have
*   destroyed it, like you want to destroy me, the analysis,*
*   everything.*
I don't want to destroy you!
No?
No! No!
Good. Because I'm still here.

10

Crisis

Sam was a friend of a friend. He was ten years older than me, handsome and affable, but also insecure, a teller of bad jokes – not my type. But I was way beyond choosiness. I told myself that I liked him, then that I loved him, then that I couldn't live without him. We became lovers in the early spring of 1985 and by the summer I was desperate – terrified that he would leave me, that he would just up and go one day and I wouldn't survive his departure. Every morning I woke shivering with fear. I began persecuting him, demanding reassurances of his affection and berating him frantically for the disaster that I was sure was coming. He remained calm under fire, too calm, I later realized – a previous girlfriend had attempted suicide. V didn't like what he heard of Sam, but he didn't tell me to stop seeing him, as he had with an earlier lover of mine whose wife had tried to kill herself. 'Not a killer' was V's mordant assessment of Sam, which in the event turned out to be true, although I pushed us hard in that direction.

> *Why am I with Sam? What the hell am I doing with him? He says*
> *such dumb things.*
> *You are trying to make him hate you.*
> *Why? Why?*
> *I don't know – an alibi for suicide?*

'Everyone goes away,' I write in my journal. 'Everyone breaks a promise. Everyone leaves me in this terrible place.'

My affair with Sam was meant to end everything. I didn't want to kill him or me, but I was in a frenzy of destruction. I weaved between him and V, getting drunk at Sam's flat after sessions, telephoning V from there to leave crazed messages. I began ringing V in the middle of the night to scream my despair.

Why am I behaving like this? I'm a lunatic, I must be crazy, why am I doing this?
You are trying to make me hate you.

My upper torso filled with hot lead, I could hardly drag myself about. My terror of aloneness became unbearable. Left alone, I would die; my feelings would devour me. I could be alone only when drunk or asleep, when a swarm of malevolent figures would keep company with me. Waking, I would lie frozen in shock at the violence of these dream-images, the gaping, slavering mouths, the dismembered bodies. Who was this person, this 'I' who harboured such demons?

And why would anyone care for such a person?

Why does he stick with me? He must be crazy.
Who?
Sam, of course!
I thought maybe you were referring to your crazy analyst.
[Long silence] *Are you crazy?*
Well, I must be, mustn't I, to stick with someone as bad and crazy as you?
[Long silence] *You know . . . that's not funny.*
I'm not trying to be funny.

Sally is worried about me. I am drinking three or four bottles of vodka a week, tossing back pills like a party drunk guzzling peanuts. She comes to the house one lunchtime and finds me stupefied in bed. She washes and dresses me and carts me off to my GP.

Dr J is a kind young man; he stares gloomily at my case notes. I have been getting my tranquillizers on repeat prescription, no questions asked. 'So I've been giving you all this stuff . . .' I imagine he is angry but his voice is gentle. 'What do you want me to do with you?' All I want is to pop more pills and return to bed, but the NHS has me in its sights at last. Maybe I should be hospitalized? The suggestion is made kindly but it horrifies me. What is he talking about? I am me, not some common-or-garden sicko, and anyway, what about my psychoanalysis? Dr J, as it turns out, is keen on psychoanalysis, so he doesn't dismiss this concern out of hand. And the presence of Sally seems to convince him that immediate hospitalization isn't necessary. But something has to be done about my boozing and drugging. He sends me to a psychiatrist who wants to hospitalize me straight away. No, no – I won't do that. The psychiatrist mentions a day hospital. I have no idea what this is but I say it sounds OK. So the psychiatrist refers me on to Dr D at the Whittington Day Hospital. But when Dr D's appointment letter arrives, I panic.

> *I'm not going to see her.*
> *No?*
> *No! I've got you, why should I go to a psychiatrist?*
> *She might help you.*
> *You're supposed to be helping me!*
> *Yes, and look at you. I don't seem to be helping you much right now,*
>   *do I?*

I won't go to see Dr D: what can she do for me? I cancel the appointment. A few days later I come home to a phone message from Dr D, asking me to come and see her. 'I'd like to meet you; I hope you'll come.'

> *She left a message asking me to go see her. She says she'd like to*
>   *meet me.*
> *Did she? That's nice.*
> *I guess I better go.*
> *I guess you better.*

Dr D: Same age as me. Casual trousers and a cotton shirt, just like most of my friends. Too familiar, too much the same, too much like me. What can she do for me?

I tell her my story. I am doggedly, ruthlessly truthful; nothing is left out – not the booze, the pills, the panic, the rage. I parade myself before her in all my sordid misery. She has said she wants to meet me – who did she think she was going to meet?

She appears unfazed, and after a while my fear subsides. I like her smile.

I'm afraid of how much I'm drinking.
What do you think the drinking is for? I mean, what would you do
*   if you didn't drink?*
[This is an interesting question, and I ponder it.] *I think I might*
*   smash my head against a wall.*
Well . . . we don't want that. But I think I can help you with the
*   drinking.*
Stop me drinking?
Help you to bring it under control. If you want to do that.
Hmm . . . doing all this psychoanalysis, and then drinking so
*   much – it doesn't make much sense, does it?*
It does if the analysis hurts too much.

This strikes me as an astonishing thing for a psychiatrist to say. Then she says:

You are afraid you might destroy the analysis.
[How does she know this? This is a secret between me and V.
   How does she know?] *That's what [V] says. He says I'm trying*
*   to destroy it.*
Do you think you are?
I don't know. I don't want to.
I think I can help you with that. I can help you stay in analysis. If
*   that's what you want to do.*

So there was a bargain struck.

I came away from the Whittington with a notebook. I was to use it to record how many pills I took, how many ounces of vodka I drank. I did this on most days for several years. In the worst times, the crazily drunken times, I stopped, but I always started up again later. The notebooks make for hard reading now, but they helped at the time.

Sam and I lurched on into the autumn. Christmas approached: V would be taking his usual fortnight break; I was crazed with pain and fear. In the week leading up to the break, as I approached V's building in the mornings, a terrible joy welled up in me. I trembled ecstatically as the hair rose on my arms and the back of my neck. Laughter bubbled into my throat like vomit.

What is it, this feeling? It's a terrible feeling! What is this feeling? It's the joy of destruction.

As I left my final session before the break, V gave me his holiday telephone numbers.

The next morning Sam and I travelled to Wales, to a cottage near the coast. He brought comedy videos, card games, a bird to roast. I brought my anaesthetics: vodka, tranx and liquid valerian. Sam must have known how much I was drinking, but he said nothing. The first couple of days were manageable – we walked, ate, watched television. I kept myself topped up with sedation; maybe I could keep the lid on? Little was said, and when I tried to snuggle up to Sam in bed he froze against me. By the middle of the week my nights had become wild. The pain was excruciating, and the Laughing Woman was euphoric. Sam had moved into some unreachable zone. He went for long runs (he was an exercise junkie), made cinnamon toast, chuckled over his videos. I huddled into a corner, watching a knife disappear between his shoulder blades. By Christmas Day I was so full of drugs and booze that I could hardly move. Loathing and terror shuddered through me. Sam surely couldn't take much more. 'You hate me, you hate me!! Don't leave me, don't leave me!!!' 'Barbara, don't beg me! I can't stand this!' I tottered to bed with a full bottle of vodka; the air had gone thin and white.

V was right: Sam was no killer. At some point during the night he took the booze and pills from my bedside. In the morning he got me out of bed and held me while I vomited. 'I must ring V,' I told him. I could barely walk. Sam supported me to the phone box up the road.

I feel like I'm going to die here. I've got to get out of here.
It sounds awful.
What's happening to me?
I think it's a repetition. It sounds awful.
I'll go back to London; I'll go to Cora's. Sam will drive me there.
Ring me when you get there.

Sam took me to Cora's and drove off. Cora fed me some soup and put me to bed. Later I rang V and he set up telephone sessions with me for the remainder of the break.

Should I go to hospital? I can't go on like this.
That is worth considering.
What should I do? What do you think I should do? I can't go on like this.
I can't decide for you. You must decide what you need to do.
How can I decide? I don't know what to do. What should I do?
You want me to tell you what to do so that you can prove how useless I am. Your plan is for me to tell you what to do so that you can go off and do it, and it works out very badly. You will make sure that it works out badly to prove that my advice was wrong.
I won't do that!
No? How have you ended up in this situation then?
I don't know what to do! You're supposed to be taking care of me! What should I do?
I'm your psychoanalyst, not your mother — I don't run your life for you. You must decide what to do. Ring me on Wednesday at noon.

What was I going to do? How could I survive this? I rang Dr D's office but she was away. I felt overwhelmed by helplessness. The crisis had come and no one was looking after me; I was going to die.

Meanwhile people all around me were exerting themselves to help me. V interrupted his holiday to take my calls – sometimes several a day. Cora installed me in her spare room, locked up her alcohol and kept an eye on me through the rest of the analytic break. Friends rang up, came round, took me with them to Christmas parties where I slumped in a corner. Cora bore the main burden. Her workdays were disrupted, she cancelled social events to stay with me. I felt no gratitude. Whatever Cora felt she kept to herself. One night, prowling the apartment looking for alcohol, I found a half-bottle of Pernod that had escaped the cull. I drank it all and was horribly ill. I felt ashamed and guilty, but also recriminatory. 'Why didn't you lock it up with the rest of the booze!?' This was a bit rich, even for Cora. 'Barbara, I am not your mother!' But she was wrong. For me, then, everyone was my mother. And anyway, what were my female friends for, if not to look after me? Otherwise I might as well stick to men.

Women

I have been a feminist most of my life. I was just finishing high school when the first reports of Women's Liberation trickled into Saskatoon. Its advent came as a shock to me, as it did to most girls growing up in the 1960s. The times may have been a-changing, with civil rights, Vietnam, flower power blooming and wilting, but women's lot seemed writ in stone. The canons of femininity may have been less strictly policed than in Mum's youth, but they were still depressingly rigid. Open servility to men was no longer demanded, at least not in the middle-class world where I grew up, but subtler forms of deference were still expected. 'Real women' walked a tightrope between ditzy girlishness and 'mannishness'. 'Career girls' like Mum had to be cautious. When Mum was appointed to the bench, the local newspaper made much of her family responsibilities. Later she dyed her hair and this was duly reported under the headline 'Judge changes her mind'. I was twelve and found this very charming.

So the change, when it came, found me wholly unprepared. 'What do you think of feminism?' a friendly boy asked me in 1968. 'Oh I believe in it!' I told him, 'I think women should be very feminine.' He didn't laugh but lent me Simone de Beauvoir's *The Second Sex*. Working in Vancouver the following summer, I picked up inky little pamphlets proclaiming the power of sisterhood. I listened to bright young women with body hair denounce left-wing men (like my father) who expected women to service them. Soon I became a convert, then an enthusiast. My family came out to Vancouver at the end of the summer and I told them about my discovery. This immediately launched me into terrible fights with Dad, who trotted out old-left indictments of 'bourgeois feminism'. A meal in a Vancouver restaurant ended in an argument between Dad and me over equal pay that was so protracted and savage it drove Lise from the restaurant in tears.

In 1971, a few months before I left Saskatoon for London, a well-known English feminist came to the local university to give a lecture. I was part of the group that invited her, and I brought her home for dinner. She was blonde, gorgeous, eloquent. She terrified Dad. I watched him watching her, as she expatiated about abortion reform over the dinner table, and his expression shamed and excited me. How could he let such a woman intimidate him? And how could I become such a woman?

London in the 1970s, I soon discovered, buzzed with such women. No anxieties about cleverness here. Everywhere I went, I encountered women who handled ideas with an aplomb that thrilled me. A few years earlier such women, when they left university, moved off into families or careers or, like Mum, into some tricky combination of the two. Now they gathered under the Women's Liberation banner and turned themselves into a campaigning intelligentsia. This became my gang, my home. All my friends belonged to this world. With them, I studied Marx, Freud, the political economy of women. I wrote papers, organized conferences, taught courses in women's history, helped to edit a socialist-feminist journal with the in-your-face title of *Red Rag*. At the peak of the Women's Liberation Movement, in the mid-1970s, I attended a meeting almost every day, sometimes two or three. In 1979 I spent twenty-four Thursday evenings in a shabby women's centre in Islington, testing out my first book, chapter by chapter, on a class of feminist activists, who gave it the thumbs-up.

This was a vibrant, creative milieu, but it was not a gentle one. Women's Liberation was a very angry radicalism. Eventually its fury overwhelmed it. There is a wild frustration that can afflict social movements like feminism, whose equalitarian premises are so obviously true that their sheer banality can be maddening. It was enraging to find oneself arguing with beauty-contest defenders and 'pro-lifers' and other sexual reactionaries, and that rage became toxic, seeping into our relationships with each other. We dreamed of a radical unity of women, a political sisterhood, and for a short time this seemed almost possible. But there were too many inequalities among us – of class, ethnicity, cultural advantages, financial resources – for a lasting solidarity; and anyway many of us were much angrier with our mothers than with any man. Activists turned on each other, criticizing, carping, guilt-tripping, until finally the movement imploded. Many important initiatives survived, and the feminist impress on western intellectual culture deepened throughout the 1980s. But by the time Britain acquired its first female Prime Minister, Women's Liberation had disintegrated.

My friendship circle survived the meltdown. We all belonged to the same left wing of the movement, and we were enough alike – mostly white, middle class, university-educated – to cope with our differences. We took our feminism into colleges, universities, publishing houses, teaching and writing about women, editing journals with a feminist bias. With my friend Sally, also a historian, I taught adult education classes in women's history, Marxist-feminist theory, psychoanalytic feminism. I was excited by the launch of the feminist press, Virago, and delighted when its editors agreed to publish *Eve and the New Jerusalem*. And when I broke down, shortly after *Eve* was published, it was my female network (along with a few close male friends) that held me as I descended into illness, and continued to hold me until daily life became impossible and I was hospitalized.

When politicians talk about 'community care' what they really mean is women: women inside and outside families; women struggling, often with meagre resources, to look after loved ones who are too crazy or old or physically incapacitated to look after themselves. I was lucky. My friends were mostly well resourced and imbued with ideals of personal solidarity. They shared me out among themselves, occasionally consulting with each other to ensure that no one took sole responsibility for me. I had genuine 'community care', unlike the imaginary kind to which people with mental disorders are mostly consigned these days. But even that was not enough. 'I can't do this any more, Barbara,' Charlotte said to me one afternoon, a few weeks before I went into Friern Hospital. I had been drinking steadily, and that afternoon I had nearly fallen down her front steps. Now I was hunched up, shaking, on the end of her sofa. 'I don't think your friends should try to do it any more. You need professional nursing.' I looked at her, looking at me sombrely, and huddled down deeper into myself, hating her.

Gratitude is a social emotion. It needs at least two; no one can be grateful alone. It also needs freedom: we do not feel grateful to people for doing things we compel them to do. So it can be a difficult feeling to access, especially for people – like infants, and like me at this time – who believe that their feelings control everyone around them. Before I became very ill I had appeared to be an adult, even a competent and successful one. But my grown-upness had been fictitious. Now I was showing my true colours. I was a baby-emperor, ruthlessly dictating to my friends. My need for support was real enough but it was also an alibi, a way of reassuring myself of my omnipotence. After my Christmas crisis with Sam, I spent day after day on the phone organizing care rotas for myself. Friends who hesitated to commit were given short shrift. Often I would double-book to insure against cancellations. The inconvenience this caused was of no concern to me – wasn't my life at stake?

My friends were very concerned and so tolerated the bullying, more or less. V was less accommodating. Session after session he battered at my grandiosity. He spoke when he chose to speak; he came and went without my permission; he said forbidden things. When I phoned him, he sometimes kept me on the line longer than I thought he should . . . or shut me up mid-tirade. He might do anything! It was intolerable. If I was not all-powerful I was completely helpless. I rallied furiously. Unable to rein him in, I comforted myself by ringing my friends endlessly to organize and reorganize arrangements, treating them as bit players in my inner drama. One night I dreamed I was locked in an empty department store among draped mannequins – a scene of utter desolation. V was unsympathetic:

But that is your life, isn't it? Alone among people who have no life except that which you breathe into them.

I go home and write in my journal: 'It's true. I have no one. I can't go on like this. Will I have to go into hospital?'

Bad Dreaming

I never saw Sam again. He phoned a few times but I didn't speak to him. I was too ashamed of how I'd behaved . . . and I had had enough of men.

I have had enough of men.
Oh yes?
I can't do it any more.
Do what?
Sex and love . . . and that.
What's that?
What's what?
You said you'd had enough of that – what did you mean? The penis?
Oh that! I don't know. I don't know what I want. Anyway, I've had enough of it.
Enough penises?
Oh for God's sake!
Enough sex? Enough love?
I don't like sex. And I don't know anything about love.
No.
No what?
I don't think you know about love.
[Silence] *But I love you.*

This sudden declaration from me shocks me very much.

The crisis through which I had just passed had left me in a quandary. What was I going to do? How was I going to spend my days? My ruthlessness with Cora had shaken me slightly. I knew I had been pushing my friends very hard and I feared alienating them. But I had to have company; I would disappear without it. Maybe some voluntary work was the answer? I knew my friends would be relieved, and if I had to be with other people all the time, I might as well be doing something worthwhile.

My first job was with a women's group that campaigned on feminist issues in the Third World. The women, who were mostly Latin American, were puzzled by my request for unpaid work, but they were also pleasant and kind. They gave me stacks of press clippings to file. But their tiny office was bitterly cold and I left after a fortnight.

The next job took me to the Feminist Library, a small collection of Women's Liberation literature on the second floor of a Georgian house near the Embankment. The two young women who ran the library put me to work cataloguing and answering the phone. *Eve and the New Jerusalem* sat on the shelves. One morning I took a call from a prominent historian whom I knew slightly. I gave her the information she requested and hung up without identifying myself. At the end of the day I took *Eve* from the shelf and shoved it into my bag, alongside the quarter-bottle of vodka I always carried with me. I never returned the book.

The Feminist Library occupied three of my weekday afternoons; the remaining two I spent in a left-wing publishing house, helping out with this and that. Weekends I volunteered at a residence for the physically disabled. This was a wonderful place. The dozen or so inhabitants were severely disabled, with cerebral palsy, spina bifida, multiple sclerosis, but they ran their own lives. Everything they needed done they did for themselves, by proxy; the staff, including me, acted as their feet, hands, eyes. All decisions belonged to the residents; we helpers had to do just as we were told. There must have been some limits to this, but I never witnessed any. The results could be messy. Cooking sessions were often chaotic, especially when directed by the younger residents, whose favourite ingredients seemed to be chocolate and ketchup. Pub outings sometimes ended in sick drunkenness. I cleaned rooms, pushed wheelchairs, changed colostomy bags, wrote letters. I got a van licence so that I could drive people around. I felt suicidal much of the time, but sometimes I also felt useful.

Two of the women in the residence were regular churchgoers. Edna was a quiet, elderly lady who loved going to Sunday services at St Paul's Cathedral. I would wheel her up to communion, ahead of the rest of the congregation, feeling fraudulent. The other woman, Myra, was a black American who belonged to a Pentecostal congregation. I was enchanted by her church, with its music and shouting and home-made cakes, and longed to belong there, if only to lay my head on the broad bosom of a Jamaican woman sitting next to me, roaring her devotions. One Sunday, when I was feeling very low, I told Myra a little about me. Later that afternoon, as we left the church after service, she touched my hand. 'Jesus looks on you, Barbara, he sees your sufferings.' I brushed her away, stiff with misery and humiliation.

I did these jobs for about five months, until I could no longer manage them. I could barely look after myself, much less anyone else, and the everyday civility the work required was beyond me. I was a sick and terrified baby; I needed care. 'What can I do? Where can I go? What can I do?' I scrawl in my journal. 'How am I to go on? Who will be with me?' I drink heavily, swilling down pills; the Laughing Woman whoops with delight.

At night when I fall asleep, I go straight to hell. My dream-world is an inferno; some nights I vomit with horror and fear.

Dreams did not feature much in my life before psychoanalysis. 'I never dream,' I told friends confidently. But after I started seeing V

I quickly became a dedicated dreamer, especially after I began recording my dreams on waking. Within a year I was recording three or four dreams a night, many of them long, elaborate narratives. I began my sessions by recounting these to V. 'You are bringing me gifts,' he said. But our discussions of them didn't seem to go anywhere. The 'royal road to the unconscious', Freud had said of dreams, and my unconscious obliged by producing reams of symbol-laden dreams most of which, however, led into dead ends. 'What does that make you think of?' V would ask of some minor detail – the colour of a coat, the number on a door – and I would draw a blank. And why was he wasting time with such trifles anyway, as day after day I relayed to him seduction scenes with Dad, penises in all shapes and sizes . . . 'Very Freudian,' he would remark drily of these. Well, they were: so why didn't he appreciate them?

It didn't occur to me that dreams could prevaricate or lie. And free association – that lodestar of the psychoanalytic method – completely eluded me. Asked about some element in a dream, I would rack my brains . . . I had always been a good student, so why were my efforts here so fruitless? V would sigh, and our session would move on.

The truth was that I didn't see the point of free association. V had the dream dictionary, the interpretive key. He was the doctor with the know-how; I was the patient just waiting for him to pronounce. A leading psychoanalyst, Thomas Freeman, once joked to another analyst that 'the difference between ourselves and organically based psychiatrists is that they know what is going on. We haven't the faintest idea, so we have to listen to our patients to see if they can help us!'[1] Had I heard this joke at the time, it would have seemed no joke to me. I didn't want to have to explain myself to V. I had no notion of collaborating with him; or rather, I paid lip service to this core element of the psychoanalytic method while undermining it at every turn. I was caught. Part of me knew all too well what my dreams threatened to expose, and I twisted away from this knowledge, dredging up bits of this and that to chuck at V as a sop. Sometimes, alarmingly, these fragments led somewhere. V would make a suggestion and I would feel it hit home. Later I would fill with pain, or have a panic attack. Then I would retaliate by dreaming a string of convoluted dreams whose sole aim was to prove V's stupidity. He would take the bait, the interpretation would come along, and then – nothing. Gotcha! 'You are very determined not to learn from me!' – and I was, because what he might teach me terrified me beyond words.

In 1983, a year after I started analysis, I published a children's book titled *The Man Who Stole Dreams* (a film was later made of it which frightened me when I first watched it). The book tells the story of a town whose inhabitants are being robbed of their dreams by a man who boxes them up and sells them back to the townspeople at a 'dream supermarket'. Two children recognize their dreams in the supermarket and demand their return. When the man refuses, the townsfolk, led by the children, release their nightmares from the boxes. The nightmares overcome the man and the people return to their beds with their rescued dreams. The nightmares are left behind, swirling around the sleeping head of the vanquished thief.

Now, three years on, all my dreams have vanished except the nightmares, and the man to whom I give them, day after wretched day, seems incapable of doing anything with them. My despair is total, and I am sure that I will not survive much longer.

15 March 1986

Three dreams last night – can I bear to write them down?
– A baby falls to the ground and splits apart at the mouth. Dad impales the little body on a stick and waves it in front of me. The baby's guts spill out; Dad laughs and smears the guts across my face.
– A baby lies on its back in a pram, trussed up like a chicken about to be roasted. Its eyes are wild with terror. A voice says: 'This baby is going mad with fear.'

– A little girl has lost her family. She has a new family, but she can't find it. The girl wanders about helplessly. She comes to a field that is covered in ice. A woman is standing on the other side. The child walks across the field to the woman. 'Are you my new mother?' The woman laughs contemptuously. 'Not me – I'm dead.' She pulls open her coat and shows the girl her torso, which is ripped open. Her belly has a baby's head inside it. The girl turns to run away but the woman grabs on to her, yelling, 'What have you done to the baby?'

Why am I filled with such horrors? How have I come to this hell? Where is my Virgil?

All my life, somebody else has told the story of my life. Now there is no one. 'There is nobody who writes this,' I write in my journal. 'I am a pounding heart, a dazed emptiness, a shrill hatred, an absolute loneliness, a life frozen into pieces. I have no idea who writes here. I cannot understand why I have not died, except that the fact of being alive is a hard fact to change.'

13

Inferno

What do you want from me?

What do you mean?

I mean, you come here every day, just like clockwork. You bring your dreams, we talk, you go away. And how are you now? You tell me you are feeling worse. You drink, you take drugs, your health is suffering. Coming here is not helping you, it seems. It seems to be making you very ill.

You're not making me ill!

No, well, how do you feel? Do you feel any better? Your life has become very difficult, in some ways even more difficult than it was before. How does that make you feel?

What do you want me to say?! I feel awful! But it's not your fault!

No, whose fault is it then?

I don't know! Mine, my parents' . . .

Your parents! Where are your parents? I don't see them here. There's just you and me here. And I am your psychoanalyst, and I'm supposed to be helping you to understand yourself, so that you will feel better.

Aren't you doing that?

Am I? Is that what I am doing? Is that what you want from me?

Of course it is! [Is it? Is that what I'm here for? Is that what I want from him?]

Psychoanalysis works, when it works, by transferring the patient's 'illness' on to the analytic relationship so that its mechanics — the unconscious fantasies and fears that drive the illness — can be exposed and, over time, dismantled. This is not a genteel process. I was evil incarnate, a creature of supreme malignancy; only a fool would take me on. After four years of analysis, V's stupidity was an article of faith for me. Perhaps he was competent with other patients — what did I care about that? — but not with me; my badness had swamped him, unmanned him, melted his brain. My worsening state was testimony to this. The alternative — that I was helplessly dependent on a person over whom I had no control — was too terrifying to contemplate. Any intimation of V's separateness from me produced overwhelming fear and rage, and yet the reality of his otherness from me, the sheer fact of it, pressed on me constantly. 'Don't touch me!' I screamed once, as he leaned forward slightly and I heard his chair creak behind me. He never touched me physically, not even a handshake. But his words tore into me, so that I would sometimes jerk on impact.

In the winter of 1986 I journeyed through a sequence of phobias. Old phobias reappeared and new ones gripped me. A childhood fear of dogs returned and I avoided parks. A man brought his spaniel into a newsagents while I was buying a newspaper and I rushed out without the paper. I had never experienced agoraphobia, but now open doors frightened and confused me. Going up and down the stairs at home became difficult, and I couldn't enter supermarkets. Fears seemed to proliferate uncontrollably. My face-washing accelerated and I began bathing at night, desperate to rinse away my filthy dreams. Some years earlier, when I had been very anxious about an analytic break, V had given me a small pebble. I had kept the pebble with me all through that break and through subsequent ones. Now I took the pebble to bed with me and slept with it clenched in my hand. It gave me some comfort, but my night-times were torture.

Occasionally comprehension broke through. A run of frantic days and horrible nights finally ended when a dream jostled me into recognizing what I had been seeing, but refusing to see, for some time: V was no longer wearing a wedding ring. I could not believe that such a thing could cause such suffering. My furious incredulity left him unmoved. 'You blame yourself for whatever you believe has happened to my wife.' His wife! He had (but no longer had . . .) a wife! I was stunned that he had used the word. Then a wave of excited delight broke over me – he has left her because of me! – followed by a sharp spasm of guilt: what has happened to her? Why have they split up? Has he really ditched her because of me? He must have ditched her because of me! How terrible! Now he must hate me! What will he do to me?? Suddenly I heard my thoughts. Oh.

Oh. Yeah, OK, OK. Yes, I see what you mean. Yes. Good.

That night I slept well for the first time in months.

Another patient of V, a smartly dressed man in his forties, approached me in the hallway one morning. I had just arrived for my session; he had just come out of his. He smiled at me as if he knew me. I was a patient of V, wasn't I? Would I like to have a coffee with him sometime, maybe next week? I mumbled something and fled to the toilet. When I reported this to V he was mildly dismayed. 'You want me to promise you that this man won't approach you again.' By now I was elated – what fun! Another patient wants to take me out! – but later in the day I began to shake with agitation. For days my heart galloped as if it were trying to race out of my chest. When I picked up my pen to write in my journal, the pen seemed to caress my genitals in a teasing, sickening motion. I thrilled with disgust. The Laughing Woman hooted and hollered with delight. 'Cut off my breasts,' I wrote. 'Chop off my clitoris, wrench away the skin from my face which so disgusts me. I am dead, dying, dead, I died somewhere a long time ago. This hell is where I went. Leave only clean bone . . .' I drank heavily. 'Alcohol is my mother now,' I told my journal. My mouth filled with spittle and I saw myself spewing up a thick black liquid that bubbled and fructified. My nights filled with images of sexual mutilation, of vomit and blood and shit. One night I watched a man's head being devoured as he spoke. I lugged this image dutifully to V the next morning along with all my other nocturnal horrors. He was not appreciative.

There's nothing new here.

What?

This dream, there's nothing new here. There's nothing new in any of these dreams. All I am hearing is the same thing — how much you hate me, how much the analysis is harming you . . .

What? These dreams? They're not about you and the analysis! They're about . . . I don't know . . . I don't know . . . sadism, I guess.

Yes, they are certainly sadistic! They are horrible. They are showing us how horrible your mind is, how hopeless it is to try to change you.

Hopeless! Do you think this is hopeless!? What are you saying??

I'm saying that's what your dreams are telling us. I don't know if it's hopeless, how should I know? That's what your dreams say. And they say that I'm a fool to think that I can help you.

You . . . ? Oh my God! It's hopeless, my mind . . . I can't stand it any more. What am I going to do? What am I going to do?

Leaving this session, I turn at the door and look back at V. The sombreness of his expression terrifies me.

And then, in the midst of all this, we have a friendly chat about my love of history. I have discovered that William Beckford, a notorious Georgian libertine, probably homosexual, once lived in this house. But when I report this to V he says only a polite 'Really?' I am surprised: doesn't he find this fascinating? 'I am interested in other kinds of histories,' he says, slightly defensively. Then: 'Not everyone has your interest in British history, I mean a professional interest.' I push aside this 'professional' – what are the chances of my ever researching and teaching history again? – and we spend the rest of the session talking about why I became a historian, about the 1970s and the rise of women's history, about my involvement with the *History Workshop Journal*. I am still attending editorial meetings for the journal now and then, and I do a bit of writing and editing. V is pleased about this, and I am pleased that he is pleased. But the happy intermission ends on a sour note: 'I can't imagine ever writing another book again. My book about Wollstonecraft . . . I don't even know where the files are . . .' I sigh with pitiless complacency. V bites back: 'Oh? Thrown them away, have you?'

In February 1988 V moved his consulting room from Harley Street to West Hampstead. Instead of a large, elegant building I now find myself in a small Edwardian house on a dull suburban street. No leather chairs, no receptionist, no copies of *Country Life*. The waiting area is just a little L-shaped alcove with a bench and a cactus plant. I panic: what has happened? Is V in trouble? Has his incompetence been exposed? For a few days I see no other patients in the house and my fear intensifies. What are V and I doing here all alone? Eventually more therapists and patients appear and my anxiety subsides, but I remain convinced that the move is a downhill slide for which I am somehow responsible. My punishment must surely come soon . . .

And then the punishment comes, although not from V. I am going to lose my home. My housemates cannot live with me any more; I must sell them my quarter-share of our house and go. If I don't agree to this, they will put the house on the market. They know that I won't force them to sell, and they believe that the money from my share of the house will allow me to buy a place of my own. In financial terms, they are right. I must buy myself a flat and move out.

They don't want to live with me!
No . . .

How can they do this to me? It's my home too! They don't want to
live with me!
You *don't want to live with you.*
What? . . . [Long silence] *Yes, that's true. That is true.*

So what to do with me? In the days and weeks that followed I oscil-
lated between a murderous rage, fuelled by booze, and a fugitive
glimmering of new possibility. I lay on the couch execrating myself for
having alienated my housemates and threatening suicide. V seemed
bored; I wasn't sure he was even listening any more. Then one morning, at the start of a session, I felt my face twist into a wolfish mask, all
teeth and clotted blood; the horror transferred to my belly, where the
thing crouched and clawed until I was gutted, rolled up, kicked away . . .
The images, almost hallucinatory, transfixed me; then –
'This is just me,' I said to V. 'Doing this to me. Just me. Me myself.'
'Yes.' A deep peace fell on me, and when I returned to Highbury I
wrote in my journal: 'I think maybe I do want a place of my own instead
of always hiding, hiding in rooms in other people's houses – never my
own – something of my own – I have not even thought I could wish
for it . . .' I began looking at small apartments and one day I saw one I
thought I could afford. I returned home, planning my move, and woke
the next morning rigid with fear. I was ill; I had not been able to be
alone for nearly three years; how was I going to do this? I doubled up
in terror and began drinking, and I didn't stop drinking until I was
admitted to Friern Hospital ten days later.

I go into Friern on a Thursday. My analytic session that morning has
been a blur of agonized rage. (V's rage, as well as my own: 'You drove
here in this condition!' he almost shouts at me. 'What the hell do you
think you're doing?') In the afternoon I stagger to an appointment
with Dr D, who takes one look at me and rings the hospital. 'Maybe
we could see how I get on over the weekend . . . ?' I protest feebly.
'Would you make it through the weekend?' That silences me.

Cora drives me to Friern, her face set. After leaving Dr D I had gone home and drunk more vodka, and I arrive at the ward so sozzled I can hardly remember my name, much less the name of the recently exploded North Sea oil platform which the admitting psychiatrist, to test my condition, asks me about. 'How have you gotten in this state?' he inquires the next day, but I have nothing to say. What on earth *can* I say? So I am given pills for alcohol withdrawal and packed off to bed, where I lie quivering, my brain blipping on and off, until drugs and exhaustion push me into sleep.

I have been to Friern once before. The psychiatrist who referred me to Dr D three years earlier has her office there. On this earlier occasion Sally comes with me, to make sure I don't fudge things. 'You must tell the truth, Barbara.' She imagines this will ensure effective treatment.

The Friern psychiatrist, Sally says after our visit, 'is a sadistic bitch'; I think maybe she is a witch. She is old and very white: white hair, white coat, dusty white skin. The insides of her lips are a dry orangey-red; I can't stop staring at them. She is the gingerbread witch; I keep my fingers well away from her. I don't want to answer her questions; but I am, after all, a good girl, so I rattle out my story with Sally watching me closely. The witch wants to hospitalize me straight away. 'You cannot do psychoanalysis in this state. It is absolute nonsense. You must come off the alcohol and drugs. You are killing yourself.' I will not agree. I do not realize that she can 'section' me, that is, legally force my hospitalization, but for whatever reason she decides against this and refers me on to Dr D. Let her look after this foolish woman! 'You are a very sick girl.' She scribbles for a while, then leans across her desk to hand me the referral letter. I snatch the letter and lurch away, away from the dried-blood lips, the cold whiteness.

Back in the hospital reception lodge, I collapse on to a bench. A man comes over, begging for cigarettes; Sally shoos him off. I look around at the peeling paint, the dirty windows, the ancient, scabby brick. The reception desk is a mahogany cubbyhole with a broken ivory bell. It's a bloody museum, I think; someone should be selling tickets. Sally reads out to me the inscription on the 1849 wall plaque commemorating the laying of the foundation stone, on which the asylum is lauded as the 'pride and boast of the county'. I am unimpressed. 'What a hellhole.'

PART TWO

14

The Asylum

Friern was not just any loony bin. When my friend Raphael Samuel visited me there, what he saw with his historian's eye was not the sad, doomed place Friern had become, but what it had once been: Colney Hatch, the emblematic institution of the Asylum Age. Entering the hospital in July 1988, I became part of a ghostly lunatic army, one of the tens of thousands of people who had resided there – some for days, some for decades – since its founding a century and a half earlier. When the hospital closed in 1993, my patient records travelled to the London Metropolitan Archives along with the rest of the hospital's files, which today occupy more than eighty linear metres of shelving at the LMA. Perhaps one day some future chronicler of Friern will take my records down from the shelf and peruse them alongside those of my fellow Friernites, and so see for herself something of the story I am about to tell here, about the world in which I found myself in the twilight days of this famous old asylum.

By the time I was admitted to Friern – drunk, sick, suicidal – I had lost all sense of myself as a historian. My life had imploded and I had collapsed inward; nothing outside me mattered, least of all the old dump I had landed myself in. But as I dried out my mood lifted, so that by the time Raphael came to visit me, a week or so later, I was curious enough about my surroundings to question him. 'So, what was this place then?' 'Barbara, darling! Colney Hatch! Don't you know about it? You must read about it, write about it!' Raphael was a notorious enthusiast but even so his excitement impressed me, and I recalled it five years later, in the early spring of 1993, when I heard that Friern was mounting a valedictory exhibition on its history. I went along to the exhibition with my friend John and we spent a couple of hours peering at architect's plans, patient registers, photos of industrial workshops, summer fetes, the Friern football team . . . Like Raphael, John was fascinated. 'What a place, what a story! Maybe you'll write about it someday.' This seemed implausible; but then – remembering Raph, who having urged me to write about Friern was always sure I would survive to do this ('You're too tough to go under, Barbara') – I reflected that anything was possible. 'Yes, maybe.'

Many people have seen inside Friern Hospital without knowing it. A B-movie image of a loony bin, its vast decaying wards were so nightmarishly atmospheric that photographers and film companies queued up to use them.[1] Few extant interiors so faithfully mirrored the gothic inner landscapes of madness. The hospital's most famous feature – a corridor extending over a third of a mile, the longest in Europe – could make even the stoutest spirit quail. Stretching out endlessly, its vaulted roof and dirty walls striped with light from narrow windows, empty except for an occasional figure shuffling along, muttering and gesticulating, this corridor was the very emblem of despair. Friends traversing it for the first time, en route to my ward, arrived wide-eyed with dismay.

At its founding, Colney Hatch – it didn't become Friern until 1937 – was the largest asylum in Europe.[2] Opened with much fanfare in the Great Exhibition year of 1851, Colney Hatch was, in conception at least, no gloomy Bedlam but a showcase for enlightened psychiatry. Its lovely grounds and elaborate frontage – an Italianate riot of campaniles, cupolas, rustic stone quoins and ornamental trimmings – signalled a prestige institution designed to comfort and heal the truant mind. Madhouses were notorious for 'managing' their inmates with chains and whips, but now this new asylum, in quintessentially Victorian fashion, put them to work instead. Like most of the great nineteenth-century asylums, Colney Hatch was a self-supporting community. Its 165-acre site boasted a large farm, orchards, gardens, stables, gasworks, waterworks, laundries, bakeries, and craft workshops manufacturing everything from brushes and beds to boots and clothing of all varieties. Most of the asylum's food and, by the end of the nineteenth century, all of its clothing were produced on-site by the patients. Even the beer accompanying the patients' dinner (until a busybody subcommittee of the London County Council banned beer from the asylum in 1891, to much protest) was brewed in the asylum brewery.

Such industrious self-provision, especially in the year of the Great Exhibition, marked out Colney Hatch as a model Victorian institution. Many people coming to London for the Exhibition travelled up to Friern Barnet to visit the asylum, to marvel at its size and grandeur and to nod appreciatively at the sight of hundreds of lunatics labouring peaceably in its fields and workshops. A few years later such tourists could, if they wished, attend a 'lunatic ball' (fifteen of these were held in 1868 alone, along with magic-lantern exhibitions, concerts, lectures and plays) or the ever-popular summer fete.[3] The exemplary conduct of the patients at these events strongly impressed visiting guests, including one visitor to the 1869 summer fete:

> The day was one of the hottest this year. In a meadow near the asylum there had been erected several tents, at which beer, tobacco, tea, coffee, and other refreshments were freely sold at a reasonable rate. There was also a band of musicians beside conjurors, acrobats, Punch and Judy, &c. Kiss-in-the-ring and other games were carried on . . . Besides the patients and attendants present, there were county magistrates and ladies, dressed to the astonishment of the many admiring patients. There were metropolitan guardians, and such friends and relations of the patients, numbering many hundreds, as chose to visit them on this celebrated occasion . . . out of 1,200

females, over 500 were permitted to take part in the proceedings; and out of 800 men, 300 were also on the ground . . . All these, carefully selected no doubt, participated in the excitement going on . . . Of the majority it seemed all but incredible that there was any urgent need of entire seclusion from society, so much did they seem to enjoy this passing glimpse of the outward world. During the afternoon two patients, both of them apparently affected by the heat, required removal, which was morally accomplished, without any interruption of the pleasures going on.[4]

(In July 1989 I went along to the Friern summer fete. No beer was served, and there were no Punch and Judy shows, but otherwise it was much the same event as 120 years earlier.)

So idyllic did all this appear that it left more than one mid-nineteenth-century observer convinced that Colney Hatch was a model environment for the sane as well as the insane. The only concern was that patients residing in such a cheerful and healthful place would never want to leave.[5]

Yet within a few decades Colney Hatch had become a byword for neglect and misery. Studying the asylum's history, I marvelled that it lasted as long as it did.[6] Perhaps we should see its continued existence as testimony to the degraded state of public psychiatry in the years before the 'community care' revolution – certainly this is how many would interpret it. But matters are more complex than this. Friern's history exemplifies a phase of western psychiatry that began on a high tide of reformist optimism and then descended into troubled waters before finally foundering in a flood of anti-institutional, anti-welfarist sentiment. Whether its disappearance, along with the rest of the asylums, is a victory for improved mental health care is not clear. History's verdict has yet to be delivered, and it is possible that the judgement will be more favourable to the old asylums, at least in some respects, than psychiatric modernizers would like us to believe.

Lunatic asylums were a Victorian invention. Before the mid-nineteenth century most lunatics lived with their families. Those without families, or whose families could not or would not look after them, roamed the country alongside paupers, vagrants and other social outliers. Religious charities provided some care, but this was limited and haphazard. In 1676 the famous Bethlem Hospital ('Bedlam'), which had succoured disturbed minds from the fourteenth century, moved into a new building in London's Moorfields district – England's first purpose-built institution for the insane. Bethlem remained the country's only public madhouse until the mid-eighteenth century, when it was joined by a handful of charity-funded madhouses in London and other urban centres. Poor lunatics sometimes lived in these institutions, but more often they ended up in parish workhouses or jails.

England also had privately owned madhouses. These were few in number until the late seventeenth century, when population growth and a weakening of community and kinship networks led to an upsurge in private madhouse-keeping. Wealthy lunatics who could not be looked after in their homes were placed in these houses, as were growing numbers of pauper lunatics whose costs were met by their parishes. By the mid-eighteenth century mad-keeping had become a highly lucrative business, with many entrepreneurs turning a hand to it. Yet even at the height of this 'lunatic trade' patient numbers were low, with most private madhouses catering to a half-dozen people or less. The few public mental hospitals were also very small by Victorian standards, housing at the most 200 or 300 patients. The situation was the same in other western nations with the exception of France, where several large lunatic asylums were founded in the seventeenth century. Georgian England had more private madhouses than other countries, but even so in 1800 there were only around 2,000 people in specialized lunatic institutions in the whole of the country.[7]

The nineteenth century saw this situation change utterly. In 1807 a House of Commons inquiry into the state of lunacy in England recommended the establishment of county asylums, financed on the poor rates, and in 1845 the provision of such asylums was made mandatory. A Lunacy Commission was created to regulate all institutions catering to the insane. Within a few decades public asylums had sprouted up at the edges of towns and cities across the country, and by the end of the century over 100,000 mentally ill people were housed in English asylums.

Similar developments occurred across most industrialised nations. State asylums began to be built in the United States in the 1840s; by 1904 there were 150,000 people living in these institutions. Canada too built its first public asylums in the mid-nineteenth century. These were small institutions, followed by much larger ones in the early twentieth century. In 1921 Saskatchewan opened a huge asylum in a small city named Weyburn. 'He's gone to Weyburn,' a school friend confided to me after her older brother vanished one day. 'What's in Weyburn?' I asked. 'Crazy people, loonies!' she giggled and ran off.

This 'Great Confinement' of the insane, as Michel Foucault famously anathematized it, was fervently endorsed by medical opinion.[8] Sequestration was vital for disordered minds, Victorian mad-doctors insisted.[9] Families reluctant to lock their loved ones away were assured that it was only confinement in a well-regulated, salubrious environment, under the care of medical specialists, that offered any hope of recovery. Home-based nursing could never match skilled asylum therapeutics. Some asylum enthusiasts went even further, arguing that home was the worst place for a lunatic 'because there circumstances that excite a maniacal paroxysm frequently exist'.[10] After all, as the anti-psychiatrist R. D. Laing was to argue over a century later, it was often the lunatic's family that had driven him or her crazy in the first place.

But the strongest case for the new asylums lay in what they had replaced. Some pre-Victorian madhouses, especially those catering to the well-to-do, had been decent institutions, but the overall picture had been horrific. Medical men responsible for the new asylum system, like John Conolly, a leading psychiatric reformer of the 1830s and 1840s, recalled all too well the terrible conditions of the recent past:

> In the gloomy mansions in which hands and feet were daily bound with straps or chains . . . all was consistently bad. The patients were a defenceless flock, at the mercy of men and women who were habitually severe, often cruel, and sometimes brutal . . . Cold apartments, beds of straw, meagre diet, scanty clothing, scanty bedding, darkness, pestilent air, sickness and suffering, and medical neglect – all these were common . . .[11]

From 1774 all madhouses had to be licensed, but the regulation was poorly enforced and reports of beatings, whippings and even starvation were legion. Public opinion had been slow to react to this. Most Georgian Britons regarded lunatics not as suffering fellow creatures but as demons, monsters, wild beasts – 'mad dogs, or ravenous wolves' as one London mad-doctor branded them.[12] Insanity was sub-human, and those afflicted by it had no place in the great chain of being. They could, however, provide good entertainment: for most of the eighteenth century Bethlem Hospital was a popular tourist destination where, for a few coppers, visitors could stroll about gawping at gibbering figures chained to walls or locked up, naked, in filthy cells.[13] Bethlem eventually came to symbolize the evils of the unreformed madhouse system, but the practices and 'treatments' employed there – shackling, bed-strapping, strait-waistcoats, bloodletting, induced vomiting and frigidity (the asylum was kept icily cold, as this was said to have sedative effects) – were near-universal in the eighteenth century. The acute suffering this caused was barely acknowledged, as most medical authorities held that lunatics were insensible to physical discomfort: 'All mad folks in general bear hunger, cold, and . . . all bodily inconveniences, with surprising ease.'[14]

Such attitudes were not universal however, and by the mid-eighteenth century there were signs of change. Enlightened physicians argued that mental disorders were brain diseases, not signs of diabolic influence or bestial states. Novelists and poets moved people to tears with images of minds unhinged by grief, lovesickness and other emotional travails. Madness was gradually humanized: lunatics were not brutes, it was said, but sick souls deserving of succour. In 1770 Bedlam was closed to tourists, and in the following decade King George III's episodes of derangement earned him much sympathy.[15] But the most important change came with the development of a new therapeutics known as 'moral treatment', which by the end of the eighteenth century was exerting a strong influence on institutional psychiatry.

Moral treatment was a portmanteau term for therapeutics directed at the minds and emotions of lunatics rather than any supposed organic cause of insanity. At a minimum, moral treatment required asylum-keepers to manage their charges without recourse to 'mechanical restraints' (shackling, chains, et cetera) or corporal punishment. In its stronger versions, it meant the abandonment of the ineffectual and often brutal 'medical' remedies popular among mad-doctors, in favour of a psychotherapeutic approach that utilized the asylum environment and staff–patient relationships as healing agents. In 1795 a French doctor named Phillipe Pinel was appointed physician-in-chief at the Salpêtrière Asylum in Paris, where he instituted a system of *remèdes moraux* based on personal rapport and clinical observation, engaging his patients in 'repeated, probing, personal conversations', taking detailed notes as they spoke.[16] Pathological ideas and emotions (the 'secrets of the heart') were identified and, where possible, gently challenged.[17] Medical treatments such as bloodletting and purging were rejected in favour of the cultivation of strong personal bonds between asylum-keepers and their patients. Psychological healing, it was argued, was an up-close process, entailing a high level of emotional involvement on the part of the would-be healer. 'To truly benefit the lunatic,' Pinel's student Jean-Étienne Esquirol declared, 'one must love him and devote oneself to him.'[18]

In the same years that Pinel was revamping Salpêtrière as a moral-treatment asylum, the Quaker philanthropist William Tuke was creating the Retreat, an asylum run along moral-treatment lines in York. Founded in 1796 and made famous in 1813 by Tuke's grandson Samuel in his 1813 *Description of the Retreat*, the York Retreat embodied long-standing Quaker traditions of group support and personal counselling. Its design was cosily domestic, with low walls, pretty gardens and comfortable furnishings. Residents were treated as members of the Quaker 'family' and encouraged to participate in the religious life of the community.[19] Chains and corporal punishment were banned, although milder forms of restraint were permitted as a last resort. As at Pinel's Salpêtrière, physical remedies were mostly discarded in favour of rational conversation (to wean patients away from mad ideas) and appeals to the patient's moral sensibility.[20] 'Judicious kindness' – *douceur* in Pinel's idiolect – was the asylum's maxim.[21]

Pinel and the Tukes were curative optimists. By the end of the eighteenth century enlightened medical opinion had reconceived madness as an ailment to which any person – even a king, as in the case of George III – might succumb. Now these moral therapists insisted that the disease was only partial, that lunatics retained intellectual and moral powers which, if properly acted upon in a supportive environment, would replace 'morbid feelings . . . [with] healthy trains of thought'.[22] All human beings, they averred, have a natural propensity to sympathize with their fellows. The task of the asylum-keeper was to reignite this innate 'benevolence' through a combination of moral example and the exercise of his own sympathetic instincts.[23] In 1839 John Conolly was appointed as the medical superintendent of Hanwell Asylum in Middlesex. He immediately banned all mechanical restraints and introduced a rule requiring employees to treat all patients, 'however violent', with 'kindness and forbearance'.[24] The creation of a calm and nurturant atmosphere was Conolly's priority. The stresses of life wreaked havoc on delicate sensibilities, he argued. Domestic life in particular was full of travails and 'excitements' that could unbalance fragile minds.[25] A well-conducted asylum would be a sanctum where chaotic minds were soothed into sanity:

> . . . calmness will come; hope will revive; satisfaction will prevail. Some unmanageable tempers, some violent or sullen patients, there must always be; but much of the violence, much of the ill-humour, almost all the disposition to meditate mischievous or fatal revenge, or self-destruction will disappear . . . and despair itself will sometimes be found to give place to cheerfulness or secure tranquillity. [The asylum is] where humanity, if anywhere on earth, shall reign supreme.[26]

Conolly's regime at Hanwell was widely publicized and highly influential. Responses from his fellow mad-doctors ranged from sceptical to openly hostile, but lay reformers and legislators were more persuadable, and by the mid-nineteenth century a dilute version of moral treatment, combining non-restraint with some traditional medical remedies, had become public policy.[27] In 1847 the Lunacy Commissioners reported very favourably on the 'substitution of mild and gentle treatment in place of the old method of mechanical coercion', and by 1854 twenty-seven of the thirty county asylums in England and Wales had abandoned mechanical restraints.[28] Visitors to the Hanwell Asylum were delighted to witness the inmates gardening, attending chapel and even dancing at a Christmas party without 'a single circumstance occurring to mar [their] happiness'.[29]

These happy times were the high point of the new Asylum Age. But they were short-lived. In the second half of the nineteenth century asylum populations rose rapidly, as pauper lunatics crowded in from the workhouses, and wards 'silted up' with the 'chronically crazy'.[30] Moral treatment foundered under the combined pressures of

overcrowding, 'cheeseparing economies, overworked medical superintendents . . . untrained, under-supervised nursing staff'.[31] By the
late 1860s most asylums had reintroduced straitjacketing and other
physical restraints.[32] North America saw a similar pattern of decline
as early asylums founded on moral-treatment principles deteriorated
into repressive, nightmarish institutions, objects of fear to their neighbouring populations.[33] By the end of the nineteenth century the
curative confidence of the asylum pioneers had vanished entirely, to
be replaced by a hereditarian determinism as gloomy as the decaying
buildings housing the 'degenerates' and 'defectives' that lunatics had
now become.[34] Care collapsed into custodialism, as the mad were
pronounced 'tainted persons', and the asylums became their
prisons.[35]

The story of Colney Hatch Asylum – *my* asylum, as I still think
of it – exemplifies this sad history. Colney Hatch opened at the height
of the moral-treatment boom. 'No hand or foot' would be bound at
the new asylum, the chairman of the Middlesex magistrates declared
at the stone-laying ceremony, for here was no mere jail but 'a curative
institution . . . and we anticipate that with the advantages which this
asylum can command, it will soon acquire a European reputation'.[36]
Yet even before the construction work began, John Conolly was
warning the Middlesex officials that Colney Hatch's projected size of
1,000-plus inmates would militate against the 'close and intimate'
care needed by lunatics. Wards would be unvisited, patients neglected,
and 'many good principles . . . hopelessly given up'.[37] And sure
enough, within a year of the asylum's opening, the Lunacy Commissioners were expressing concern about overcrowding and insanitation.
A few years further on, and a visitor to Colney Hatch was struck by
the contrast between the asylum's impressive facade and what lay
behind it. 'The enormous sum of money expended upon Colney
Hatch . . . prepares us for the almost palatial character of its eleva-
tion . . . [and] the whole aspect of the exterior leads the visitor to
expect an interior of commensurate pretensions.'

He no sooner crosses the threshold, however, than the scene changes. As he passes along the corridor, which runs from end to end of the building, he is oppressed with the gloom; the little light admitted by the loopholed windows is absorbed by the inky asphalt paving, and, coupled with the low vaulting of the ceiling, gives a stifling feeling and a sense of detention as in a prison. The staircases scarcely equal those of a workhouse . . . In the wards a similar state of things exists . . . of human interest they possess nothing. Upwards of a quarter of a million has been squandered principally upon the exterior of this building; but not a sixpence can be spared to adorn the walls within with picture, bust, or even the commonest cottage decoration . . . There is no more touching sight than to notice the manner in which the female lunatics have endeavoured to diversify the monotonous appearance of their cell-like sleeping rooms with rag dolls, bits of shell, porcelain, or bright cloth.[38]

Conditions continued to worsen. In 1861 the Lunacy Commissioners issued a highly critical report on Colney Hatch, listing among its many defects faulty ventilation, 'bedsteads with sacking but no mattresses' and the 'habit of bathing many patients in one lot of bath water'. Stung by the criticisms, the asylum's Board of Guardians responded by tartly reminding the Commissioners who the asylum was designed for: 'Colney Hatch Asylum was established for Pauper Lunatics only, and [the] many luxuries and appliances suggested by them [the commissioners] are quite unsuited to that class of patient.'[39] Nothing daunted, four years later the Commissioners returned to the attack. By now the asylum's population had doubled from its original size, mechanical restraints were in use, and the overall atmosphere, especially in the 'refractory wards', was so oppressive that it drove the Commissioners to strong words: 'It would be difficult to instance more perfect examples of what the wards of an asylum . . . should not be, than are presented here . . .'[40] Five years further on, the *Lancet* pronounced Colney Hatch 'a colossal mistake', and there were calls for the asylum's closure.[41]

The downward spiral continued relentlessly through the rest of the nineteenth century. In 1896 a temporary wooden building was erected to house infirm female patients. The Commissioners warned that the building was a fire risk but the warning was ignored and seven years later the building burned down, killing fifty-two women. The years after the First World War saw some small improvements: lockers for patients' possessions, tea served 'in a household manner' on a few wards. The asylum now had a large number of recent Jewish immigrants among its patients and a kosher kitchen was provided for them, along with a Jewish cook.[42] These were welcome innovations, but overall life at Colney Hatch remained austere at best and, in the case of chronic patients languishing on its notorious 'back wards', cruelly depriving.

In 1930 a new Mental Health Act introduced voluntary patients into the asylum system. For the first time people could enter asylums, and leave them, without compulsion. Outpatient services were also initiated, and it was assumed that asylum populations would now begin to decrease. But they did not, and in fact admissions continued to rise well into the 1950s, including admissions into Friern, which in 1954 were running at three times their 1939 rate.[43] Nevertheless, the presence of voluntary patients presented a major challenge to the custodial ethos of the asylums, a challenge that mounted steadily towards a crisis in the wake of the integration of the asylums into the National Health Service in 1948.

Initial plans for the NHS excluded psychiatric services, possibly due to lobbying by asylum doctors who feared loss of power as they competed with general hospitals for funding and status within the state system. But Aneurin Bevan, the Labour Minister of Health responsible for the introduction of the NHS, was determined to end the ghettoization of mental health care, which he regarded as 'a source of endless cruelty and neglect', and the old mental hospitals were propelled into the new regime. From this moment, although no one knew it at the time, the asylums were doomed. How could places that locked people up, subjected them to involuntary treatments, frequently neglected or even abused them, be part of a modern health system? No reform-minded government could tolerate it.

At first the asylums responded well to the challenge. The Second World War had seen some major innovations in institutional psychiatry (I say more about these in Chapter 17), and these continued to gather pace after the war, eventually cresting in a reformist wave that swept through the asylums, bringing with it new treatments and rehabilitation programmes, the unlocking of wards, and a revitalization of moral therapy.[44] David Clark, the psychoanalytically minded medical superintendent of Fulbourn Mental Hospital, was a leader in these changes, introducing psychosocial therapies in the 1950s and 1960s that soon became very influential under the banner of the 'therapeutic community' movement.[45] Assisted by the introduction of new symptom-suppressant drugs, psychiatrists in many asylums – including Friern – began experimenting with group therapy and other psychoanalytically inspired treatments, and the sector was gripped by a resurgent curative optimism.[46]

Some asylums turned their backs on these changes, but enough embraced them to bode well for the future. Yet it was at this very moment that the government, in the person of Enoch Powell, then Conservative Minister of Health, sounded the death knell for the asylum system. Addressing the 1961 conference of the National Association for Mental Health (now Mind), Powell attacked the old hospitals as medical dinosaurs. 'There they stand,' he told his audience: 'isolated, majestic, imperious, brooded over by [the] gigantic water-tower[s] . . . the asylums which our forefathers built with such immense solidity to express the notions of their day.' But this day had passed: 'For the great majority of these establishments there is no appropriate future use.'[47] Powell meant what he said: a year later he issued his Hospital Plan providing for the replacement of the mental hospitals by acute-care psychiatric wards in general hospitals and community-based services for non-acute and aftercare. The 'deinstitutionalization' of mental health care – to use the unlovely neologism coined by sociologists – or 'decarceration', as others dub it, was under way.

Powell was a Tory libertarian, but his assault on the mental hospitals earned him plaudits across the political spectrum, not just in Britain but internationally. Asylums across the western world were moving into crisis. The recent improvements in British asylums, scattered and experimental as these were, had few parallels elsewhere.[48] American mental hospitals were vast, anonymous 'bins' often housing 10,000 patients or more, most of them held there under compulsion; Canadian asylums were generally smaller but no better. In the same year as Powell delivered his 'Water Tower' speech the sociologist Erving Goffman launched an excoriating attack on the American hospitals. In sharp, vivid prose, Goffman's *Asylums* (1961) revealed the day-to-day degradations inflicted on inmates of a Washington asylum. Waving aside the hospital's medical pretensions, Goffman condemned it and its counterparts as 'human storage dumps': a judgement echoing that of Thomas Szasz who the previous year, in a hugely influential book titled *The Myth of Mental Illness*, had damned the asylums as prisons presided over by psychiatrist-jailers. Goffman's and Szasz's indictments of the asylum system were echoed by 'anti-psychiatrists' in other countries, including Michel Foucault in France, Franco Basaglia in Italy and R. D. Laing and David Cooper in Britain: a formidable band whose collective onslaught on the mental health establishment became a high point of 1960s radicalism.[49]

In Britain, Laing, Cooper and their supporters formed the left wing of an anti-asylum offensive which also included welfare reformers, investigative journalists and, eventually, patients themselves, organized into a growing 'consumer' movement, as well as Powell and his supporters.[50] The language of these campaigners was militant – but they were pushing at an open door. In 1959 a new Mental Health Act had abolished the distinction between psychiatric and general hospitals and eased voluntary access to inpatient mental health care. In its wake, the ratio of voluntary to detained patients had risen dramatically: a transformation which made the custodialism of the old asylums appear ludicrously oppressive as well as outdated.[51] Outpatient psychiatric services were also expanding, as new drug treatments made it possible to treat increasing numbers of people in their homes. By the early 1970s asylums everywhere were recording a steady shrinkage in their resident numbers.[52] Moreover, many of the buildings that housed the old hospitals were falling apart. Renovating them would be hugely costly: an unwelcome prospect to governments, especially at a time of fiscal crisis. And then there were the scandals. The 1960s and 1970s saw a steady stream of exposés of neglect and abuse, which together delivered the *coup de grâce* to the Asylum Age. 'Every few months . . . some sort of scandal is reported,' a psychiatrist lamented in *The Times*, ending his article with a plea to the government to 'finish the job and close down these old hospitals'.[53]

Friern was in no condition to withstand these shocks. It was poor (funding allocation per patient was still far below general hospitals), its physical fabric was crumbling, and its reputation was shaky. Then in 1966 it was hit by a major scandal when a campaigner for the elderly, Barbara Robb, revealed serious abuses in its treatment of old people with dementia. Robb's report, and the government inquiry that followed, led to calls for closure, although no official action was taken until over a decade later.[54] But the writing was on the wall, and Friern was running out of allies. After all, who could defend a place that handled people like 'human trash', as one journalist demanded in the wake of the government inquiry into Robb's findings.[55]

The final decades of the asylums were very difficult ones. By the late 1960s the community care revolution was gathering pace: wards were closing, leaving their former residents at the mercy of 'community services' which, in most areas, were still rudimentary or non-existent. Homelessness was skyrocketing. By the end of the 1970s the asylum population had declined by two-thirds: a truly astonishing rate of reduction, especially when one realizes that it was achieved (as one of western psychiatry's most astute critics, Peter Sedgwick, commented at the time) 'through the creation of a rhetoric of "community care facilities" whose influence over policy in hospital admission and discharge has been particularly remarkable when one considers that they do not, in the actual world, exist'.[56] Opposition MPs declared community care a 'catastrophe': a judgement later endorsed by a leading figure in UK mental health politics, Baroness Elaine Murphy, who titled her account of community care between 1962 and 1990 'The Disaster Years'.[57]

Friern's closure was announced in the midst of this debacle. In July 1983 the hospital learned its fate from a televised news announcement.[58] Staff were stunned: all those I have interviewed, including its medical director at the time and several consultants, insist they had no inkling of any closure plans. Indeed only three years earlier the Hospital Management Team had issued a glossy brochure, 'Friern 2000', celebrating the hospital's achievements and looking forward to the next millennium.[59] But the North East Thames Regional Health Authority had done its feasibility studies and added up its sums, and Friern was for the axe.

The years that followed the closure announcement were painful ones for the Friern staff, some of whom organised a robust anti-closure campaign calling itself CONCERN, which was soon outgunned by its opponents.[60] 'We were quite a militant group,' Doris Hollander, a consultant on my ward who led the campaign, later recalled, '[but] [t]here were other powerful organizations saying "It has got to happen now" . . . There was no shortage of people who could point to all the terrible things in the old hospitals and disregard their positive side.'[61] Hollander's views were aired in Parliament, where in July 1990 the Labour MP for Islington North, Jeremy Corbyn, described a state of 'panic' among patients at Friern, and demanded assurances that the hospital's closure would not proceed unless adequate accommodation for its inmates could be guaranteed.[62] A few of these patients and former patients also made their voices heard, attending meetings with Friern managers where they expressed support for the hospital's closure mixed, however, with strong concern about post-closure provision.[63] Even the local vicar got in on the act, collecting fifty signatures on an anti-closure petition. But to no avail: the decision-makers were unbudgeable, and the hospital's fate was sealed.[64]

As the clock ticked towards closure, a team of researchers moved into Friern under the aegis of a government-funded study into the impact of the hospital's closure on its decanted residents. These researchers followed the progress of patients into staffed group homes over a five-year period and reported, for the most part positively, on their lives there.[65] The people being studied however had been selected for their capability while many others, usually more disabled and so harder to place, remained in hospital; some were still there on the night before Friern closed.[66] Moreover – and more important for the long-term consequences of the hospital's closure – over a third of the decanted patients required readmission during the five-year follow-up.[67] Alternative inpatient provision was radically insufficient, and seriously ill patients entering the hospital in its twilight days faced an accelerating crisis of resources. Many of these 'new long stay' patients, as they were awkwardly dubbed, did not qualify for the new community facilities: what was to become of them, when Friern's doors closed for the final time?[68]

In 1989 I was a likely candidate for 'new long stay' status, or at least that was certainly how I saw myself when I entered Friern for the third time and remained there for nearly six months. My stints in Friern – a fortnight in July 1988, followed by four weeks in May–June 1989, and a final six months from June to November 1989 – came midway through the closure process, and the evidence of this was everywhere. Many of the nurses had left and been replaced by agency staff.[69] The ward across the stairwell from my ward was empty, having been burned out in a major fire the previous year. (Its charred remains were still clearly visible through the porthole in the ward door.) Hallways were sealed off, therapy rooms shut down, the old apple orchards were choked with weeds. The kiln in the outdoor pottery shed broke down and was not repaired; a little pot that I left in the firing queue was thrown away.

Like many on my ward, I was anxious about the impending closure. Where would I go, when Friern was no more? What would become of me? 'I'm frightened about the hospital closing down,' I said to some fellow patients one day. 'It has to close,' a man named Simon, a big fellow with a rich Scottish accent, said to me. 'The devil is here, down in the cellar. They tried getting him out, but he won't go. Can't you hear him down there?'

15

First Day

Waking up in a mental hospital isn't something you plan for. My first morning in Friern, I surface on a tide of queasy amazement. It's not what I see that astonishes me – I've sat through enough loony-bin films to have some idea what to expect – but who I now am. There have been moments in recent months when I have hardly recognized the desolate woman inhabiting my body and brain; but sooner or later the familiar self would always reappear, sporting her labels – historian, feminist, writer. Now I am in a place that redefines me. Now I am a loony, a nutter, one of those forlorn beings who lurk in the dark recesses of our society. My me has drained out of me; I am on the far side of the moon.

The memories recounted in this book have not been easy ones to retrieve. But the confusion of self that accompanied my admission to Friern makes recalling that time especially difficult. My diary entries dried up for a time: who would write them? I felt naked, stripped of my identity, my history. When I told a nurse that I had published a book, he smirked at this piece of blatant make-believe. Would I ever write another book? It seemed as improbable as flying (I dreamed often about flying at this time). I mourned the woman I thought I had been.

But over the weeks and months and years, as I moved between psychiatric institutions, my attitude changed. After all, who was it really that I had lost? The woman I had worked so hard to keep afloat – with

her sharp opinions, her self-assurance, her small triumphs – had been
a fake, a simulacrum of personal success. Now the frightened, miserable creature crouching inside her was fully exposed and I could shed
all pretence. No need to sham normalcy here. Now I was in a world
of devils and phantoms and wide-awake dreamers, an outlandish place
where my nightmare fantasies and anxieties were merely routine. This
ordinariness of my sufferings didn't make them any less gruelling, but
it lowered their temperature a bit.

Another change accompanied this. Living in an asylum is, in its
unique way, a test of character. Psychoanalysis had uncovered my
self-loathing. Digging beneath my grandiosity, I had found only garbage and shit. Now I began to rediscover my competencies. On days
when I could, I read and wrote in my journal; occasionally I even did
some editing of the *History Workshop Journal*. Performing these familiar activities in Friern gave them a new significance. My mind was
not just a cesspit. My energies, so ferociously destructive in recent
months, had their creative uses. I had thought I knew this, but now
it felt like a revelation.

Nor was I entirely a monster of egoism. I had long taken my friendships for granted. In my worst times, I had viewed my friends through
a haze of self-disgust – as worthless as I was, so were they. But mostly
I valued them for the things I thought valuable in myself: cleverness,
right-mindedness. My friends were unconventional but they were
also successful professionals, with all the usual accoutrements of their
class. Now I found myself making friends with very different kinds
of people, the sort of people I would previously have shunned, whose
day-to-day lives were much tougher than mine and whose survival
powers, as I got to know them, seriously impressed me. And as these
new friendships evolved, so did the self that I brought to them. The
Barbara of these relationships had none of the bright-girl glamour
that I had striven so hard to cultivate, but she was more open, more
adaptable, possibly kinder, than I had thought myself to be. 'Talk to
Barbara,' I heard a woman say to a new patient bewildered by ward

life – and I swelled with a pride that was new to me. Would I have forgone those long months in Friern? Absolutely. But I will never regret them.

Thin cotton curtains separated my bed from those on either side. The bed facing mine across the dormitory was empty; the room was quiet. Was I alone? A sound to the right. I looked through a gap in the curtain and saw a middle-aged woman pulling a pillowslip down over her head. A young woman in a T-shirt and jeans ran up and yanked at the pillowslip; the two women tussled over the cloth for a moment until the patient let go. The younger woman, a nurse as I now saw she was, turned away scowling – and spotted me peering out. 'You're awake. You've slept late. Better get up.' Moira, her name tag said. 'Toilets are there, bathroom's there – the duty psychiatrist will see you later.'

Moira was in her late twenties, plump, tight jeans girdling her tummy rolls. I heaved myself out of bed and discovered I was very weak. My legs wobbled and my muscles ached from vitamin depletion. But I was sober and rested – and alive. The dormitory was a long room with a very high ceiling, eight beds on each side. My bed was third from the entrance. As I headed off towards the bathroom the pillowslip woman stepped in front of me. 'Gimme a fag?' 'I don't smoke.' A disbelieving stare. 'Gimme a fag!' 'Sorry, I don't smoke.' I shouldered past her. 'Meanie! Meanie!' she yelled after me.

The bathroom had a row of sinks along one side and two big tubs in the centre of the floor. As I ran my bath a chemical cloud rose from the water, so pungent it made my stomach heave. The door had a huge keyhole but no key so I dipped in and out of the water very fast, scrambling to dry and dress before pillowslip woman came in search of her cigarette. As I left the room I hurried past the discoloured mirror hanging over the sinks, averting my eyes as I went.

Back in the dormitory, Moira was pulling the covers off a body. A hand snaked out to pull the covers back. 'No lying about,' Moira said to the body – and to me, in case I had any ideas about retreating to bed. So I picked up my shoulder bag (where did people put such things?) and went out to look around.

Ward 16 proved to be a vast open space with a nursing station at its centre. The front end, by the entrance, was empty except for a ping-pong table (no ball or paddles) and a few chairs. This was, I later discovered, the 'activity area'. At the far end were the dayroom and dining room, with a kitchen to one side of the dining room. Between these were the male and female dormitories, each with a bathroom and toilets (with unlockable doors), and a half-dozen small rooms: a couple of offices, an examination room, a meds closet and two little 'seclusion' rooms, which I later discovered had once been padded cells. As I took this in, I looked around for the people who were meant to be occupying all these spaces. The communal areas were so big, I expected to see hordes of patients somewhere. But all that was visible were a few people sitting in the dayroom – where was everyone? At its peak in the early 1950s Friern had a population of nearly 3,000; now, in the run-up to closure, it was down to about 800. This ward, designed for fifty-plus patients, now housed twenty-seven. Patients and staff rattled around like pebbles in a vacuum flask.

Ward 16 was an 'acute admissions' ward, meaning it was meant for people in a highly disturbed state, needing plenty of care (and drugs) until they stabilized and could be discharged or moved to a rehab or long-stay ward. In fact some people had been living there for years. About a third of the patients were 'sectioned' (legally detained), so the door was usually locked. Getting out wasn't always easy. You had to find a willing nurse with a key (the keys were the Victorian originals: huge iron winders straight out of a Hollywood gothic). I often spent many minutes in search of a nurse, who would then tell me to wait. 'What's your hurry? Don't you like the company here?' Getting in wasn't straightforward either. 'Ring the bell,' I would tell my friends when they asked for directions. The bell was easy to identify as somebody had scrawled *Ring Bell!* next to it, along with a further instruction to *Fuck Off to Hell!* But visitors who rang the bell – having made their way along the interminable corridor, with its bleak graffiti and muttering spectres, and clambered up the dark winding staircase to the ward landing – would often find themselves waiting for many minutes until someone came. One of my friends, upset by the long wait and overwhelmed by the weirdness of it all, burst into tears when she finally saw me.

But for now I was just an anxious new girl. I wandered about slowly . . . Not too bad, not too bad . . . What to do next? Maybe there was somebody I could talk to? I sidled into the dayroom and sat down beside a middle-aged Indian man. He was smoking with great concentration. Eventually he glanced over at me and nodded.

'You arrived yesterday.' His accent was educated.

'Yes, last night.' His trousers were an ancient pinstripe, fastened at the waist with a safety pin; his greyed vest was heavily dusted with cigarette ash.

'Could you spare a cigarette?'

'Sorry. I don't smoke.'

'Ah. Very wise.' He slumped deeper in his chair and seemed to fall asleep. But then with an effortful grunt he pulled himself back up.

'So! What has brought you here? If you don't mind my asking.'

I didn't mind, but I didn't reply. He closed his eyes and soon I heard a light snore. Should I go? I found his proximity comforting and stayed on, listening to him breathing.

Rajat had been a psychiatrist. 'My patients drove me crazy,' he chuckled at me a few days later.

'Really?' This rather alarmed me.

'No, of course not. I am a late-onset manic-depressive.'

'Oh.' In the two years that I knew Rajat I never saw his demeanour alter a jot. Either he sat quietly in the dayroom or he shuffled back and forth in front of the nursing station, dribbling ash along his path. Occasionally, if someone sat in his preferred chair, he would growl slightly – but this was his only sign of heightened emotion.

'You are one of Dr D's patients,' he said to me now when he woke.

'Yes.'

'Excellent psychiatrist, excellent. She sometimes consults with me.'

I smiled at this but a week or so later I heard Dr D talking to Rajat about a female patient who had killed herself some years earlier. Rajat had known the woman well. 'She stopped taking her medication,' he said to Dr D. 'Yes, so sad. Did you think she would?' 'Yes, very bad case . . . very bad.' He sniffed professionally.

Now, on this first day in Friern, I clung to Rajat, following him into the lunch queue, imitating him as he collected his sausage and beans from the trolley, hurrying into the seat next to him in the dining room, following him back to the dayroom after the meal. He tolerated this, but when at teatime I rose again to follow him into the dining room he turned on me: 'Go away!' I had had my first lesson in asylum etiquette: don't push it.

After tea the duty psychiatrist called me into his office. He was Mauritian and very handsome. His ivory face was etched with delicate shadows – like a moonscape, I thought. I was nervous, afraid he would send me home.

'How are you settling in?'

'OK.' Was that sufficient? 'Swell.' This word, issuing suddenly from my Canadian childhood like a mnemonic burp, sounded odd. 'That is, I feel pretty shaky but I guess that will wear off . . .' I looked at him eagerly, ingratiatingly, wanting approval. But when he tried to find out why I had been drinking so heavily, I clammed up. I wanted to say the right thing, but what was the right thing? And he was the wrong person to hear about the bad things: that was reserved for V.

'I see that you are in psychotherapy,' he said, flipping through my notes. 'Maybe best to leave that for a while, until you're feeling stronger? You can't be doing much therapy in this state.' Panic turned me icy. Dr D had promised me! She promised me I would be able to continue my analysis! My fear impressed him. 'Yes, yes, I see,' he said, reading further on, 'yes, all right, fine, you've agreed it with her, fine.' He sighed and I could see he thought this was silly. But I was

Dr D's patient, and no one at Friern ever prevented me from going to my sessions.

As I left the psychiatrist's office a young woman waylaid me. Would I comb her hair? Her hair was blonde and very fine; parts of her skull were bald. We chatted as I combed. She is Princess Alexandria of Russia; have I heard of her? I haven't, I said, but it was nice to meet her. This was deliciously on-script, and I could hear myself telling friends about it. After we finished I drifted down to the end of the ward and eavesdropped on a group of patients talking about welfare benefits. Was I 'on the sick'? one asked me. I had no idea what this meant, so I smiled vaguely and moved away to one of the huge windows that lined the ward. A light summer rain was brightening the paths across the beautiful grounds; someone was chanting a Beatles tune behind me. I watched a blackbird bustle across a wet lawn. For the first time in months, my heart felt comfortable in my chest. This was great; this was fantastic; I was never going to leave. I had found my home.

16

In the Bin

Bin memoirs are a peculiar genre. Some are modern gothics: lurid tales of decent, healthy-minded people consigned to asylums by evil or stupid doctors. Others recount the sufferings of individuals who, while they were certainly unwell at the point when they were hospitalized, experienced asylum life as cruel and degrading. And a minority are very positive, depicting the asylum as a place of healing, a sanctuary from the madness of 'normal' life. The portrayals are generally black or white, mimicking the emotional extremes of mental illness. The writer Jenny Diski, whose mother was in Friern and who spent a short time there herself in the early 1980s, captures these polarities well in her novel *Nothing Natural*. Rachel, the novel's lead character, is admitted to Friern with acute depression. As she settles into the ward, Rachel is overcome with relief: 'As she sat there a weight seemed to lift from her; she was in a place where she could just *be* . . . someone else was going to deal with the world for her.' A nurse talks kindly to her; the place is 'quiet and peaceful'; Rachel feels safe and contained. But overnight her feelings change. Now the meals regime is an irritation, the nurses are coercive, the consultant is a 'self-important bastard' who threatens to detain her if she tries to leave. Rachel flees. I encountered divided attitudes like these many times in Friern. My own reactions at first were wildly enthusiastic: a letter to a friend described the nurses as 'unfailingly kind', the patients as 'nutty but sweet', the psychiatrists as 'wonderfully caring'. By the end of my third and final admission I was much less starry-eyed. But I never lost my sense of the hospital as a refuge from unmanageable suffering (a 'stone mother', as some describe it), however bleak the physical environment and attenuated the caregiving.

Friern was a tough place to live – much tougher for some than for others. I am an educated white woman from a middle-class background. The people around me were nearly all undereducated and poor, about a third of them were black. I may have been on the skids mentally but I had far greater economic resources and cultural capital than most of my fellow patients. These things mattered. Madness can be a real social equalizer, turning a banker into a homeless bum or a psychiatrist (as in Rajat's case) into a long-term asylum-dweller. Among the severely mentally ill, class lines blur – but they seldom vanish entirely. The few middle-class patients of my ward would often sit together; people with posh accents were occasionally teased or even threatened. I was grateful for my Canadian accent, which made it hard for people to place me. But it was money that was the real divider. I had never thought of myself as wealthy – affluent certainly, but far from rich – but here in Friern, surrounded by people grinding along on state benefits, for whom a lost 'giro' or a stolen bus pass was a disaster, I learned just how hugely privileged I was. I thought I knew everything about extreme anxiety until I witnessed the agonies of a man at risk of losing his home because of a mix-up in his 'social care' payments. Poverty is a psychological catastrophe. Anyone who thinks that madness is down to defective brain chemistry needs to look harder at the overwhelming correlation between economic deprivation and mental illness. Of course there are other forms of deprivation that drive people crazy, but living in Friern I saw first hand how poverty plays hell with people's minds. Money may be shit, as Freud sort of said, but it is food and security and social belonging too, and not having enough of it corrodes the soul.

I entered Friern voluntarily. Theoretically I could leave at any time (although if my suicidal drinking had continued I probably would have been compelled to remain). Patients who were legally detained were in a very different position. One former patient I met, who had been detained in Friern many times, described receiving forcible injections that left her bruised and traumatized. Other ex-patients describe wards crackling with tension, assembly-line shock treatments, hard-ass nurses.[1] Life on the notorious 'back wards', where elderly people with dementia and others deemed chronically ill sometimes languished for decades, could be truly horrible. 'Terrible misery in that place,' one man recalls of a back ward. 'There were people there who hadn't had a conversation in thirty years.'[2]

I was spared most of this. For me, life at Friern was hard but never unbearable. And the things that made it hard – the dingy surroundings, the incipient violence, the petty degradations (and some not so petty) – suited me; they were my just desserts; this was the life I was meant to lead. The Laughing Woman, satisfied, went quiet for a time, until psychoanalysis re-galvanized her. 'Not suffering enough?' V inquired one day after I had arrived at a session hunched up with suicidal misery.

. . . I guess not . . .
So . . . What are you going to do to make things worse?
What??
You heard me.
Nothing! I'm not going to do anything!
Oh? Well . . . we'll see.

Doing psychoanalysis from Friern felt oddly right, as if my circumstances had finally fallen into line with my wretched psyche. At first I was too ill to go. Boozing and starving had wreaked havoc with my system; my muscles ached like hell; I was weak and nauseous. And anyway, why should I go to see V? He had got me into this mess; he was the enemy. Dr D would look after me; to hell with V!

This lasted five days. On the sixth morning I took a cab from the hospital to West Hampstead. After that, I went to all my sessions, travelling by train from the little suburban station just outside the hospital. For a while I didn't tell anyone on the ward where I was going, but my toing and froing made people curious and eventually I confided in a few. Rajat professed himself sorry for me: 'Pooh, Freud! Rubbish! You pay for that? And look at you, here! Rubbish! Pay me, I'll charge you a lot less than your analyst does.' But others were interested and sometimes envious. Psychotherapy had a mystique in Friern; it meant someone giving you lots of attention, really listening to you instead of just pushing pills in your direction. Several people had had some therapy and wanted more. My friend Fiona was seeing a Ms M, travelling back and forth from the hospital to her sessions just as I did. We spent hours comparing our therapeutic progress and speculating about the home lives of V and M. Another friend, Magda, was also interested, often asking me how a session had gone. 'He sounds smart, your man,' she said to me once. 'Do you think he could help me?' Her wistfulness touched me but my hackles rose slightly. 'I don't know. He's very expensive,' I said, clutching V to me.

The charge nurse took me aside one day. 'Why are you doing this therapy? Do you really think all this talking is helping you?' Yes, yes, I told him – but help was not really on my mind. Revenge was on my mind, and violence and suffering. An overwhelming sense of loss – of my home, my work, my life – fuelled a murderous vindictiveness which found some relief in piteous self-display. *Look at me! What a sorry spectacle, me here – in a bin!* The message could not have been conveyed more clearly if I had shouted it from the rooftops. *So, here I am now, in this hellish place! So much for my bloody family, my stupid friends, my incompetent psychoanalyst – what good are any of you to me?* That my friends and my analyst did not seem inclined to blame themselves for my predicament only infuriated me further. Why was nobody taking responsibility for me? All the care that I was receiving, all the strenuous efforts being made on my behalf, were swept aside in my punitive fury. As usual, V bore the brunt of this.

I'm in the bin, you bastard! So what are we going to do now?

Psychoanalysis? That's what you come here for, isn't it? Or maybe it isn't? Maybe you just come here to punish me?

Punish you? For what??

You just said. You're in a mental hospital. And this is somehow my fault.

Yes!

Yes what?

Oh . . . oh . . . I don't know . . . ! Fault, whose fault! . . . I don't know? . . . [Angry silence] What if I can't come here any more? What if they lock the door and won't let me out?

Has anyone tried to do that?

Well, no . . . But they could! And then what? Would you come to me?

No.

Never??

No, never.

So! You bastard! You drive me crazy, I end up in hospital, maybe locked up in there, and you won't come and see me?!

I haven't 'driven you crazy'. And I'm not going to come to Friern to see you, not ever. I'm not going to give you more reasons for staying in hospital.

Huh! You don't care what happens to me!

For a while most of my sessions went like this. But eventually other feelings began to come through. Cradled in Friern's unyielding embrace, I found myself surviving emotions that out in the world had felt unendurable, so excruciating that I had had to drink or drug them away. I went to my sessions sober and I stayed sober after them. Even after I left Friern I mostly stayed sober, as I took myself off to the Whittington Day Hospital and, later, to the Pine Street Day Centre. Like Friern, these day institutions were places to lodge feelings that, if left stranded, homeless, would surely obliterate me and everyone around me. Homeless feelings are boundless; they sweep all before them. Their violence is as all-engulfing as the primeval experiences – aloneness, helplessness, total vulnerability – that power them. Some memories never lose their potency; they live on in the heartbeat, the muscles, the breath. In the weeks before I entered Friern I had felt such memories pressing on me and fought them with every self-destructive weapon I had to hand. Better boozy oblivion than the horrors of helpless infancy. Now, safeguarded and sober, the feelings were pushing through, eroding my grandiosity. My ego-empire was beginning to crack, leaving me shaking with fear and despair.

> *What am I going to do? What am I going to do? WHAT AM I*
> *GOING TO DO?*
> *You sound very frightened.*
> *What am I going to do? I can't live like this; I can't be me any more;*
> *I can't be like this; I can't survive outside hospital! How can*
> *I live outside hospital? What will happen to me? What am I*
> *going to do? What am I going to do?*
> [Silence]
> *Where are you? Where are you? Oh, what am I going to do? I can't*
> *live like this . . . I CANNOT FEEL LIKE THIS! It is impos-*
> *sible to feel these things and live! Who will help me? I want to*
> *die! Where are you? Do something for me!*
> *Do you know why you are so frightened?*
> *I'm so frightened . . . I can't think, I can't think, I can't think . . . !*
> *Who will help me? . . . I can't feel like this! I can't think! Help*
> *me!!*

'The worst feeling in the world' was how V described this naked defencelessness to me. 'People will do almost anything to avoid feeling it.'

Sleep provided no respite. Dreams in Friern had an oddly redundant quality, as if the everyday derangements of the place were replaying through the nights. My nights, which relief combined with heavy drugs had made relatively peaceful at first, soon became extravaganzas of misery and horror. Heading to bed, I found myself dreading the scenarios about to unfold, crammed with images of suffering and violence, especially sexual violence. Writing some of them down later, I was so ashamed of these dreams that I carried my journal everywhere in case someone else got hold of it and read them.

A woman has a severed penis stuck up her vagina. Blood pours from her; she wails in pain. A man grabs her by the breasts. 'Suck, suck!' she yells at him. As he lowers his mouth to suck the woman drags the penis out of her body. Now it's a baby's penis, tiny and delicate with a frill of blood around the root. The woman waves the little penis at the figure sucking at her breast who is no longer a man but a baby, all battered and bloody. 'Fagh!' The woman beats at the baby; blood flies everywhere. 'Fagh! Fagh!' The screams of revulsion rocket around me and shake me awake.

A man is fucking a woman from behind. His penis skewers her and emerges from her mouth. The woman's eyes bulge; she bites down on the penis but it is rock hard. The man appears in front of her, weeping. 'I'm sorry, I can't remember how that happened. Tell me what to do!' 'I can't tell you, you bastard!' the woman shrieks at him. She is strangling on the penis, choking to death. 'You're killing me, you bastard!'

These were terrible, but worse still were the dreams of longing.

I am sitting on a bench crying. A man and a woman sit down next to me. The man puts his hand under my skirt and caresses me. 'Don't cry, don't cry.' The warmth of his hand travels up me until I feel it in my breasts, my neck. The woman says, 'Lie down.' I snuggle down with my head on her lap; she strokes my hair while the man continues caressing me. The couple smile at each other across me. 'She's too small to cry,' the woman says. 'I am crying for her.' As she weeps her tears run down my face, and I am free from all pain and fear.

I wake from this last dream shuddering with erotic yearning and wanting to die.

I can barely bring myself to tell V these dreams.

Why I am having such horrible dreams? All this horrible, crazy stuff?

All this sexual stuff, you mean?

Sexual . . . it doesn't feel sexual . . . well, some do . . . but . . . horrible! I think I'm being infected by the hospital!

Oh? Other people's dreams getting into you?

No, well — the madness . . . getting into me. All these dreams . . .

Maybe it seems safer, having these dreams while you're in hospital?

Crazy dreams in a crazy place?

Perhaps.

Psychoanalysis is tough. People imagine it as self-indulgent — all that delicious chit-chat about oneself! — when mostly it veers between pain and tedium, with occasional flashes of humour and joy. For people with severe mental illness it can be risky. By the time I arrived in Friern my analysis had taken me to the cliff edge: without the support I received at the hospital, would I have toppled over? I will never know, just as I will never know what would have happened to me without the psychoanalysis. Suicide? A life of booze and pills and intermittent hospitalizations? Or maybe some happy miracle to rescue me from my misery? I dreamed of such miracles now and then while I was in Friern, waking into grim reality — an exquisite torture. Why did I keep on with my analysis? 'Why do I keep on?' I spat the question at V repeatedly.

Why? Is it just to torture myself?
Sometimes.
Why else? Why else would I keep on coming here? I'm just making
   myself suffer more! Why else would I keep coming here?
You have different reasons at different times.
What reasons??
Well . . . to get revenge on your parents. To get revenge on me, their
   current representative.
Ah. Yes. [I have heard this so many times, it doesn't mean
   anything.]
And because you're waiting for a miracle.
A miracle? [This is more interesting. Is he offering? Maybe
   there's something he can do for me that he hasn't done yet?]
Yes, the miracle that will make you the baby your mother really
   wanted, the sort of baby she could really love, so that she would
   look after you properly.
Oh. That miracle. [Why is there never anything new?]
And sometimes, it's because you want to know the truth.
The truth? What truth?
About what happened to you.
I know what happened to me!
Do you?
[Do I?]

In my early years with V I was infatuated with psychoanalytic
theory. I read it voraciously, talked about it incessantly, formed strong
opinions on ego-psychology, Jung, Lacan. Freud was my god, I read
him devotedly. Like the twelve-year-old me in my Mennonite phase,
I looked for signs and wonders in the sacred works. For a while I
carried a volume of Freud everywhere, tucked inside my bag like a
talisman. As the analysis became tougher I began searching out case
histories of people who had lived through something similar. Eventually I came across Marie Cardinal's *Les Mots pour le dire* (*The Words*

to Say It), the story of the psychoanalytic cure of a young Parisian mother, which was marketed as autobiography. The book begins with the Cardinal character deciding to go into analysis because she has been bleeding vaginally for three years. At her first session the analyst tells the young woman that this haemorrhaging 'doesn't interest' him, and it stops immediately. A miracle! Further analytic miracles follow, eventually releasing the narrator from her misery and delivering her into a life of creative fulfilment. What joy! What a great analyst she must have had! I read and reread the book, even though I half suspected what Cardinal later admitted, that it was mostly fiction.

By the time I arrived in Friern I had lost all interest in psychoanalytic stories, fictional or clinical. Most of the time I couldn't read, and when I could I read detective fiction. Psychoanalysis no longer gripped my imagination; it was just my miserable life, my inexorable fate.

Yet on better days, on days when I could read and think, I was bored. A few weeks into my final, longest admission, I went across to the public library on Friern Barnet High Road. More detective fiction. 'I'm bored,' I told my friend Magda. The hospital had an excellent library, Magda told me; why didn't I take a look at it? 'It's a lovely old library, very quiet, beautiful rugs.' I found this pleasing image hard to conjure. Was Magda making it up? 'I'll take you there. You'll like it.'

The next day we made the expedition. I was amazed. Wood-panelled walls, Persian rugs, row after row of psychiatry, psychology, including Freud in the *Standard Edition*, all twenty-four volumes . . . The young librarian kept a watchful eye on us. 'Ask her if you can come here to read them!' Well, why not? The librarian looked at me uncomfortably. 'Which ward are you on? I'll ask about it and let you know.'

The message came a few days later. No, unfortunately I couldn't use the library. It was for staff only; if she let me use it, she'd have to let all the patients use it. Magda snorted. 'Tell her what you've done, tell her about your book.' I sent a note and waited. I think Dr D was probably consulted. A second message came from the librarian, this time welcoming me to the library on any weekday afternoon between two and four o'clock. Result! In fact I seldom felt well enough to go there, but I remember a few quiet afternoons sitting in a sunny corner with back issues of the *International Journal of Psychoanalysis* – recollections that have become sunnier over time I suspect, as they contrast so sharply with my other memories of Friern.

A friend came to visit me who was training to be a psychotherapist. She was working here at Friern, she told me, at Halliwick House, a psychodynamic unit in a separate building on the hospital grounds. I hadn't heard of this unit, but I have since met several people who were patients there. It treated mental disorders of every kind, from acute anxiety and depression to schizophrenia. I was curious to know who she was working with there. She mentioned some names and I looked them up in the library. One of Halliwick's consultants was Michael Conran, a psychoanalyst who devoted most of his career to inpatient treatment of people with psychosis and other serious illnesses. Poking around on the library shelves, I found an article by Conran where he mentioned treating people like me, who had been hospitalized during a crisis in their psychoanalysis.[3] Such crises arose, he said, when an analysis uncovered thoughts and feelings so unbearable that the patient couldn't consciously experience them and so relegated them to a 'part of the self' that then had to be looked after in hospital. 'The baby part,' I said to V after I had told him about this article. 'Yes, if the baby is desperate enough.'

A psychoanalytic patient who ends up in hospital can find herself in real difficulty. Michael Conran had seen it happen: 'The analyst may lose his patient to another doctor working in the hospital [and] psychoanalyst and hospital may then find themselves at odds with each other.' When I was admitted to Friern, V told me he was concerned that the 'hospital system' might take me over. He had been a hospital psychiatrist himself, so he knew the hazards. Part of me wanted this – maybe I could finally be rid of him! – but Dr D had no intention of letting it occur. 'You will want to continue your analysis,' she had said to me firmly when I was first admitted. On my third admission she wrote to V asking him his views on my progress. He told me about her letter. 'What did you reply?' I was afraid of his answer. 'I told her I think there's still scope for improvement.' I almost laughed at this, until a sudden surge of relief stopped me.

17

Psychoanalysis and Psychiatry

Before I began writing this book I assumed, like most people, that psychoanalysis is useful, if at all, in treating neurotic symptoms – anxiety, depression, the existential miseries of the 'worried well' – and not serious mental ailments. I thought my case was unusual. One day I said something like this to my friend Adam Phillips, a psychoanalyst. 'Well . . .' Adam said, 'actually, psychoanalysis has always treated people with so-called psychoses, as well as serious emotional illnesses like yours.' He mentioned some names and places. 'And has anyone written a history of this?' I waited expectantly. There is always a book somewhere that can tell you what you want to know. I have always believed this, and often waste a good deal of time searching for this Book. 'Not really, no. It's an important history, somebody should tell it.'[1]

It *is* an important history, one that deserves a detailed telling, well beyond the scope of this book. But since it is a history with much relevance to my story, I offer here a preliminary version of it.

Britain, with its passion for 'hard facts' and famous stiff upper lips, hardly seems a natural home for psychoanalysis. And indeed, mainstream British psychiatry, like most of western psychiatry, is biomedical: the brain is the 'organ of mind' it is said, wherein all psychological phenomena originate. This physicalism has dominated psychiatry since its inception. Early-nineteenth-century moral treatment posed a significant challenge to it, but psychiatrists soon hit back, filling medical journals with articles on 'cerebral pathology' and attacking psychological theories of mental derangement as 'absurd in the extreme'. 'It is not our business, it is not in our power to explain psychologically the origin and nature of any of the depraved instincts manifested in . . . insanity,' Henry Maudsley, Britain's pre-eminent Victorian psychiatrist, declared in 1874. 'The explanation, when it comes, will not come from the mental, but from the physical side.'[2]

The introduction of psychoanalysis into Britain at the end of the nineteenth century did little to shake this orthodoxy. Some psychiatrists were interested and a few became converts, but most dismissed the Freudian unconscious, with its roots in repressed sexual fantasy, as bogus and morally odious. Such 'secret introspective dramas' were 'more than half factitious', it was said; and even if such unconscious memories *did* exist, better to leave them undisturbed than to ferret out such 'verminous reminiscences'.[3] But while psychiatrists held Freud at bay, British intellectuals were intrigued – an interest heightened during the 1920s and 1930s by a steady influx of psychoanalytic émigrés, including Freud himself in 1938. Many writers, artists and academics undertook personal analyses and some became analysts, aided by Freud's policy – adopted by the British Society for Psychoanalysis on its formation in 1919 – of encouraging people without medical backgrounds to train for the new profession.[4]

Psychoanalysis made its first incursion into the British public health system during the First World War, when it was used to treat soldiers suffering from 'shell shock'. After the war, psychoanalytic treatment of the war neuroses continued in several new institutions, including the Tavistock Clinic, founded in 1920. In 1929, after some strenuous opposition, psychoanalysis was recognized by the British Medical Association as an appropriate treatment for neurotic disorders. By the late 1930s both the Tavistock Clinic and the Institute of Psychoanalysis (the training arm of the British Society for Psychoanalysis) had well-developed research and training programmes. Tavistock clinicians were among a group of psychiatrists recruited by Churchill during the Second World War to run psychiatric units for the military.

Yet despite this heightened profile, publicly funded psychotherapy remained scarce in Britain until the inception of the National Health Service in 1948. The early years of the NHS saw a significant expansion in psychotherapeutic services, mostly directed at poor mothers and children as part of a drive to raise health standards in the working-class population. Psychotherapeutic units opened in some hospitals and consultant psychotherapists were appointed across the NHS. But these were few in number compared to general psychiatrists, and far short of what was needed to meet an ever-growing demand for psychoanalytic psychotherapy. Priority was given to child therapy, which had received a big fillip during the war from therapeutic work with child evacuees. By 1969 up to 50 per cent of children referred to psychiatric social workers were receiving psychoanalytic psychotherapy through the NHS.[5] Peter Wilson is a leading child psychotherapist who trained in the late 1960s. 'It was a good time for child psychotherapy,' he said to me. 'When it came to kids, we therapists felt we were really making a difference.'

By contrast, adult psychotherapy in post-war Britain received little public funding (the exception to this, as we shall see, were some in-patient psychotherapy programmes). London was the psychoanalytic heartland, with only a few small outposts in other major cities. Anyone living away from these urban centres who wanted a full psychoanalysis (that is, five sessions per week with all the accompanying paraphernalia of the couch, the fifty-minute hour, et cetera) was usually out of luck.[6] But if psychoanalysis in this form was (and is) mostly confined to the metropolitan middle class, by the 1950s other types of psychoanalytic psychotherapy were having a real impact on public mental health services, including the asylums.

The main varieties of psychoanalysis practised in twentieth-century

Britain were a melange of imported and native species devised by a handful of post-Freudian innovators, of whom the most influential were Freud's daughter Anna and the superbly iconoclastic Melanie Klein, both Austrian émigrés, and, among native-born psychoanalysts, John Bowlby and Donald Winnicott. In the 1940s Anna Freud and Melanie Klein clashed fiercely over new directions for psychoanalytic theory. From the Kleinian side of this conflict there eventually emerged a 'British School' of psychoanalysis clustered around Klein, Bowlby, Winnicott and a few more influential theorist-practitioners (Ronald Fairbairn, Marion Milner, Michael Balint), which today goes by the clumsy soubriquet of 'object-relations theory'.

Despite the singular noun, object relations is not one theory but a set of theories whose common motif is the centrality of relationships to unconscious mental life. Where Freud and his early followers had focused on psychophysical drives as the motor of psychic change, object-relations theorists emphasize the role of 'significant others', especially mothers, in shaping individual psychology. The initial proving ground for these ideas was child development. Observing infants and toddlers interacting with those around them, especially their mothers, Klein, Winnicott et al. postulated an inner world of other-referring images, the so-called objects, each of which was seen to play a vital part in the drama of human maturation. These internal objects were not representations of actual people but psychic inventions, unconscious images moulded by infantile fantasy and desire. The connection between the image and the person imaged – again, the main figure here was the mother – was conceived in a variety of ways. For Melanie Klein, the connection was thin almost to the point of non-existence; the Kleinian object-image is a fantasy with little impress from reality. By contrast, for Donald Winnicott the child's direct experience of others, especially his mother, was a vital influence on his inner object-world. 'There is no such thing as a baby,' Winnicott once remarked, meaning that an infant is not a being-in-itself, but a relational entity, a being-with-someone-else, right from the beginning.[7] The Winnicottian baby is a social creature, reliant for its emotional health on 'good-enough' mothering (or if that fails, as in my case, on a good-enough psychoanalyst capable of making up the deficit).

To an outsider, these theoretical differences between Kleinians, Winnicottians, et cetera, can seem trivial. To insiders, however, they matter enormously and fierce disputes erupt, although less often today than in the past. But the shared emphasis on self–other relationships as the core of human subjectivity, combined with the famous British pragmatism, keeps schism at bay. All British School psychoanalysts, whatever their theoretical allegiances, regard the analytic relationship as the primary instrument of therapeutic change, the *mise en scène* for those revelatory re-encounters with internal others that – not always, and always with difficulty – release the self from debilitating illusions. It is through the 'transference' and 'counter-transference', chilly labels for impassioned processes, that enlightenment comes, with all its painful discoveries and startling joys.

We've seen an earlier version of this model of psychological healing before, in the moral-treatment tradition with its emphasis on the curative power of sympathetic relationships (British psychoanalysts working in the asylum system often described themselves as renascent moral therapists). Other precedents also exist, including many varieties of magical and religious healing rituals.[8] Mainstream psychiatry set its face against such precedents. In its long struggle to achieve medical respectability, psychiatry embraced scientific objectivity as a hallmark of professional identity. Like a physician tending a broken limb, the mender of broken minds was perceived as a dispassionate dispenser of expert remedies rather than as a partner in a shared therapeutic enterprise. One of psychiatry's strongest objections to nascent psychoanalysis was that it violated the 'decent reserve' of the doctor–patient relationship, opening the door to charlatanism and/or even gross abuse – of which all the 'psy' professions, including psychoanalysis, have certainly had their share.[9] But psychiatry's ethos of scientific detachment has proven no guarantee against quackery or patient maltreatment – witness the cruel handling of many asylum residents in the name of medical science.

In 1962 a leading psychiatrist tackled this issue of professional detachment head on. Writing after a long career as an asylum super-intendent, Denis Martin of Claybury Hospital in Essex criticized the injunction against emotional involvement between staff and patient, insisting that such involvement was both 'natural and inevitable'. 'Indeed there can be no meaningful human relationship into which feelings do not enter.' The answer was not to deny such feelings but 'to gain increasing awareness and understanding of them in order to use the personal relationship more helpfully'.[10] Martin was not alone in this view. The 1950s and 1960s saw a new mood in the asylums as a cohort of influential post-war psychiatrists, many of them psycho-analytically trained, set out to humanize the mental health system. Their inspiration lay in a new type of military psychiatry. As during the First World War, when psychoanalytic treatment of shell-shock victims had marked out a new path for the treatment of neurotic disorders, so during the Second World War the treatment of trauma-tized soldiers had taken the novel form of group therapy in hospitals organized as 'therapeutic communities'. Between 1942 and 1945, psy-choanalytic clinicians at Northfield Hospital in Birmingham and Mill Hill Hospital in north London cared for military casualties in wards run as self-governing communities.[11] Conventional psychiatric treat-ments were supplemented or replaced by group therapy. Formal roles and traditional authority structures were abandoned in favour of demo-cratic decision-making and a new spirit of equality and openness. As Tom Main, one of the Northfield analysts, summarized the new ethos:

> The doctor . . . no longer owns 'his' patients. Patients . . . are no longer his captive children . . . but have sincere adult roles to play, and are free to reach for responsibilities and opinions concerning the community of which they are a part.[12]

After the war, Main became the director of the Cassel Hospital (now the only remaining residential psychotherapeutic hospital in Britain), while another therapeutic-community pioneer, Maxwell S. Jones, established a community at Belmont Hospital, later the Henderson Hospital, in south London. Inspired by Main and Jones, medical superintendents in other mental hospitals – including Denis Martin at Claybury, David Clark at Fulbourn Hospital in Cambridgeshire, T. P. Rees at Warlingham Park Hospital in Croydon – began to revamp their hospitals along therapeutic-community lines. In 1953 the World Health Organization issued a report recommending the model's adoption in all psychiatric hospitals; three years later Rees, who had co-authored the WHO report, used his presidential address to the Royal Medico-Psychological Association (the forerunner of the Royal College of Psychiatrists) to urge all UK asylums to become therapeutic communities, describing this as a much-needed 'return to moral treatment' in the manner of Tuke and Conolly.[13]

Rees's version of the therapeutic-community asylum comprised open wards, occupational and social therapies, rehabilitation programmes, and the replacement of a custodial culture with one based on compassion and respect: all vital reforms, which were felt by most asylum superintendents to be extremely radical. But some, like Denis Martin, took things much further, arguing that to be effective, psychological treatment had to be an up-close, collaborative enterprise. 'There is a constructive sense in which we are called upon to "suffer with" our patients in psychiatric work,' Martin wrote; a process which 'takes a long time and often involves considerable upset and personal suffering'.[14] Martin's memoir of Claybury (*Adventure in Psychiatry*) provides a moving account of this shift from controlling patients to empathizing with them, and the huge challenges it posed to the hospital's medical staff. Interactions that breached traditional staff–patient barriers could be deeply disquieting – as David Clark also discovered when he reorganized Fulbourn Hospital as a therapeutic community in the late 1950s:

It was more satisfying to intelligent and sensitive staff and was more humane and dignified. However, it . . . was perplexing and exhausting work. To open oneself fully to the tortured feelings of the deeply mentally ill is very disturbing.[15]

Most asylum superintendents were not prepared to take on such a challenge. Nonetheless, as with moral treatment a century earlier, by the late 1950s a modified version of the therapeutic-community model had become de rigueur throughout the asylum system. Mental hospitals everywhere – including Friern – had begun unlocking their wards, establishing occupational therapy and rehabilitation programmes, and initiating regular 'community meetings'. Asylums had always been gender divided, with female patients nursed by women on one side and male patients by men on the other. The two sexes would meet only at asylum dances and other supervised social events. In the late 1950s the therapeutic-community asylums began ending this segregation. Other asylums soon followed suit and within a decade many mental hospitals had integrated some of their wards and nursing teams. This change, where it occurred, dramatically reduced the prison-like atmosphere of these asylums while at the same time creating a host of new problems. Friern established mixed-gender admission wards in 1968. 'It was a big thing, it made the place much more homey,' one consultant told me. But what about the sexual hazards, especially for women? 'Well, we didn't think so much about that at the time . . . it was the 1960s, the sexual revolution and all that, nobody thought much about the downside then.'

A radical age brought bold innovations, which occasionally extended into a rejection of the asylum system *tout court* in favour of a 'democratic community psychiatry' run by patients for patients. But this radical moment was brief. By the early 1980s therapeutic communities, inside and outside the asylums, were in rapid decline. Psychodynamic units were disappearing, and hospitals that had been famous for their community ethos, such as Claybury, were facing closure. The retreat was marked by a continuing flow of therapeutic-community rhetoric but this (according to Maxwell Jones, speaking to a *Guardian* reporter in 1984) was mostly window-dressing: 'For orthodox psychiatry it [the therapeutic community ideal] has provided a name to be wheeled out whenever it wants to defend Britain's reputation as the country which pioneered social psychiatry and to be conveniently forgotten otherwise.'[16] Some therapeutic communities, including Jones's own Henderson Hospital (now closed), survived into the 1990s, but the general outlook, in these final years of the asylum system, was very gloomy.[17]

A good deal has been written about these developments, but little by those who experienced them at the sharp end. Today this is changing, as service users rewrite psychiatric history from their own perspective.[18] But this revisionary work is still in its early stages, and the day-to-day world of the asylum resident remains shrouded in ignorance and prejudice. The ignorance is curable but the prejudice may prove more intractable. Like all conventional medicine, medical psychiatry is mostly a doing-to regimen, positioning the patient as a passive recipient of medical care rather than as an active participant in his or her treatment. The inevitable result is a denigrated image of the patient, thinly disguised today by talk of 'clients' and 'patient-led' treatment. The rhetorical shift is not insignificant, in some cases signalling real attempts at change. Yet in a 2012 survey of service users, many complained about the attitudes of mental health professionals. 'They [the staff] just say: "Ah, they're all nutters." You have staff telling you that today, you know.'[19]

The therapeutic-community movement, with its anti-authoritarian ethos and reliance on peer-based group therapies, posed a direct challenge to this view of the mental patient. Looking back on his career at Fulbourn, David Clark recalled the change:

When we started to apply the suggestions of the WHO report in 1954 [recommending the establishment of therapeutic communities]

we still thought of patients as mostly passive – people to be got working and 'activated', people to be treated or cured. We – the nurses,
doctors and planners – saw ourselves as the active ones. However,
we had gradually to revise this view when it became clear that, given
the chance to work, patients displayed surprising capacity, and given
the chance to run the affairs of a ward, they showed good sense and
responsibility. Gradually, our notions about how we should organize
the hospital began to change. I had hitherto accepted the prevailing
medical view of patients as pathetic beings, only kept from recovery
by the failure of their illnesses to respond to medical treatment or
their wilful inability to do what doctors prescribed for them. It was
several years before I even began to consider the possibility that
patients could actually help each other – and that there might be
patients who could help others better than doctors could.[20]

This acknowledgement of patients' capabilities, especially their
social capabilities, was replicated at Claybury Hospital, where the
residents, freed from rigid hospital protocols, startled the staff with
their friendly concern for one another. Even the most difficult patients
elicited 'a great deal of positive understanding' from their peers.[21] In
an open, respectful environment, empathy grew and people changed.
A woman who was at Claybury in the 1950s later recalled how emo-
tionally withdrawn she had been when she arrived, and how her fellow
patients had barged through this: '[I]t gets difficult to stay indifferent
to people who blast you with questions, affection, aggression and
problems like your own':

> It's a troublesome moment when you see that other patients care
> about you . . . You don't want the responsibility and yet need them
> to care. Straddling emotions like these can be overwhelmingly difficult
> when you experience them for the first time. On several occasions . . .
> I had an inane desire to rush back to the lounge in case people there
> had somehow forgotten I existed.[22]

I blushed when I read this passage, sitting in the British Library a few years ago. I remembered too well moments like the ones described, when I had hurried back to V's consulting room after a session, convinced that he had already forgotten about me. 'You think I cannot hold you in my mind,' he would say. 'No wonder you feel so lonely.' I was lonely at Friern too, but at least there I shared this remorseless aloneness with others. Mental illness is often destructive of social relationships, but it can also work positively, drawing people together through mutual need and a shared comprehension of the pains and confusions of madness. At the end of the 1990s fifty former asylum residents were interviewed for an oral history project.[23] Many of those interviewed spoke of the friendships they had formed in hospital. 'That's the funny thing about a psychiatric hospital,' a professional woman who was hospitalized repeatedly in the 1970s and 1980s told the interviewer, 'you make friends with people you would not necessarily come across in your normal everyday life and it's something quite unique . . . you're there in this shared experience and you see each other at your most vulnerable and there's no pretence.'[24]

The solitariness of mental illness – its 'polar privacy' to borrow Emily Dickinson's phrase – is terrible, but it is not, except among the acutely deranged, absolute. Understanding and friendship are possible, and this can make all the difference. I had long been a friendship snob, priding myself on the cleverness and sophistication of my social network. Now for day-to-day companionship I had only the hospital. The prospect frightened me. How lonely was I going to be? 'You need people around you,' V had said to me some months before I was hospitalized. But surely he hadn't meant loonies? ('Like me' – came the startling thought.)

18

Friendship

Friendship and mental illness aren't meant to go together. Before the Second World War most UK mental hospitals had a policy of breaking up patient friendships by moving one of the friends to an inaccessible ward.[1] The policy seems to have lapsed after the war but attitudes to friendship in most asylums, including Friern, remained cool. The official rationale for this was concern about potential exploitation of vulnerable patients, but the real motives were rather different. The service-user activist Peter Campbell recalls being lambasted by a nurse in the 1980s for his friendly support of other patients.[2] When I asked Campbell why the nurse had been so angry he told me the staff thought such relationships made patients 'uppity'. When I repeated this to a woman who had been a charge nurse at Friern, she pointed out to me that friendship between patients threatened the us–them divide between staff and patients. 'Friendship is too normal, it makes the patients seem just like us.'

I interviewed a man who had been the medical superintendent of a big asylum in the 1970s. He assured me that 'real friendships' were rarely found on psychiatric wards. We had been chatting for a couple of hours and I think he had forgotten why I was there. 'These people are all losers really, aren't they?' he smiled at me. 'I mean, they are people of low intelligence, low capability . . . they don't form proper attachments.'

Severalls Hospital in Essex closed in 1997. A few years before it closed the poet Diana Gittins (whose mother died in Severalls) went there to interview patients and staff for a book she was writing (*Madness in Its Place*). The patients regaled her with stories of friendships of which the hospital managers apparently knew or cared nothing. One old woman, who had spent sixty-five years in psychiatric institutions, told Gittins about her terrible grief at the loss of a long-standing female friend. The friend had been discharged without anyone bothering to inform the old woman that she was leaving. She was given no address, nor did anyone suggest that she might meet up with her friend at some future date. This indifference was repeated on a larger scale in the early 1980s, when the hospital carried out a patient reassessment exercise that involved moving large numbers of people out of the long-stay wards. People who had lived together for decades were separated. 'And a lot of people died,' a nurse told Gittins. 'They just died, because their friends were gone.'[3]

I met Fiona on my third day in Friern. She was only in her twenties, but she was a veteran of asylum life. Seeing me fumbling my way around the ward, she decided to take me in hand. Our friendship, which blossomed quickly, transformed my experience of the hospital, to the point where at moments I almost found myself enjoying my peculiar new life.

I could never figure out what was wrong with Fiona. She was a freelance designer who seemed to do well at her work. But every few years she would 'go funny' and end up in hospital. In these funny periods, she told me, she would sometimes go out late at night and get raped – this had happened three times. The rapes made her suicidal, she said, showing me the heavy cuts on her wrists and arms. Listening to her stories, I gradually pieced together a picture of a girl wandering the night-time streets, confused, distraught, available for anything. Even here in hospital, Fiona exuded a vulnerability that was somehow provocative; one day she giggled noisily at something I said and I felt a sudden urge to strike her, which horrified me. The men on the ward watched her carefully, sensing trouble. Yet she was also vigorous and funny. We spent most afternoons together, chatting in the dayroom or strolling around the grounds. One day we walked to a little parade of shops near the hospital where Fiona persuaded me to buy some pretty shoes, which I wore for many years.

I told Fiona all about myself – my friends, my politics, my book. I thought that she didn't really believe me, at least not about the book. But later I realized that she neither believed nor disbelieved. To Fiona's mind, everyone was entitled to her or his reality – me with my leftie ideas and my book just as much as her boyfriend, the 'Spaceman' as she called him, with his ten hearts and his regular journeys to Saturn, or Fiona herself for that matter, with her terrible tales of lunatic parents who regularly beat her, the men who raped her. One day Fiona took me to meet the Spaceman, a gentle Afro-Caribbean whom she had met in Friern on a previous admission. I listened as she chatted with him about his interplanetary adventures, and I realized that I didn't care whether the stories Fiona told me were true. For the truth of Fiona was not in her words but in her suffering, whatever its causes, and that suffering I never doubted, although I never saw her at her most wretched – *that* she saved for the dark streets, for the hungry, predatory men.

She did once witness me *in extremis*. I had just received a further ultimatum from my housemates. I must arrange the sale of my house share to them straight away, or they would put the house on the market. Fiona found me frantic with rage and fear. 'Whoah, that's rough,' she said. 'Come on.' She marched me down to the hospital's social work department and insisted that someone see me. The social worker was sympathetic and efficient and within a few weeks I had received the promise of a room in a psychiatric hostel. 'You just need to know the ropes,' Fiona told me as I hugged her in gratitude.

Fiona and I watched daytime soaps on the ward telly, collected windfalls in the hospital's overgrown apple orchards, ate strawberries on my bed until a nurse told us off – small pleasures that in an asylum become survival stratagems. On my second admission I met Magda, the blanketed lump I had seen the nurse trying to rouse on my first morning in Friern. Very soon we were fast friends. 'Get up! Come and talk to me!' I would plead to Magda. 'I'm lonely, I can't stand it out there – you have to get up and keep me company.' Magda suffered terribly from black depression yet nearly always she would pull herself together to be with me. Usually I did the same for her. The obligations of friendship trumped madness – and this in itself could be a form of healing.

Anyone reading scholarly studies of asylum life will guess nothing of this. Films and novels set in mental hospitals occasionally portray close friendships, but researchers either cannot or will not recognize their existence. Back in the 1950s an academic fashion arose for quantifying personal relationships through a method known as 'sociometry' or 'companion measures'. Researchers observed people whose interactions were enumerated and then displayed on graphs known as 'sociograms'. The method was widely applied to asylum inmates, including by the team of social scientists hired to study Friern during the closure process (the Team for the Assessment of Psychiatric Services – TAPS). One of the TAPS researchers undertook a numerical analysis of patients' conversations (an 'intimacy index'). Two more positioned themselves in the Willow Club, a patients' coffee bar, to observe social transactions. I knew this coffee bar – most patients who were permitted to leave their wards went there at one time or another. Day after day the researchers would arrive at the Willow Club and settle themselves into a corner table as 'unobtrusively as possible'. They would then fish out their notepads and 'tune in' to life around them. The resulting report makes for unintentionally hilarious reading. Patients are given pseudonyms (Mr Cool, Bad Teeth, Pebble Lens, Waggle Tongue) and categorized according to their behaviour patterns (Helpers, Friends, Nuisances, the Chronic). Interactions are recorded in minute detail ('Pebble Lens asks Laughter for a light – No. Hopalong is pestering for 10p. Pebble Lens gets light from Mr Cool. Hopalong gets a light from Pebble Lens's cigarette'). There are pages of this. At no point does either researcher ask any of the patients how they actually *feel* about the other people there. Nor do they ask any of the patients whether they mind being observed in this way – although occasionally they get reactions anyway: 'Coke Man discusses my writing with me: "Is it a book? Be careful not to injure anyone verbally," he tells me.' Reading this report now, I am irresistibly reminded of a pair of David Attenborough wannabes spying on a group of chimpanzees – except that Attenborough's capacity to relate to chimps seems rather stronger than these researchers' ability to empathize with the patients they are scrutinizing.

Such failures of empathy between the notionally sane and the officially mad are unsurprising. The need for companionship is a human universal but it takes many forms. When I first came to Friern I was very irritated by the constant begging for cigarettes. Since I didn't smoke, it was a while before I realized these were sometimes social overtures. By my second admission I was a smoker too and beggars now would occasionally sit next to me while they puffed on the fag I had given to them. If they offered one to me later – which happened now and then – they would often want me to smoke it straight away. If I ignored these surreptitious acts of friendliness, they would just shrug and turn away – why risk overt rejection?

I was also disoriented by the low-key aggression, often quite jokey, that I encountered at first. Should I be frightened? The answer sometimes was yes, but more often this behaviour too was a social gambit. One man who thought himself a journalist (his brother was a *Financial Times* correspondent) followed me around teasing me abrasively about my looks, my speech, my walk. Finally I sat down with him for a chat and the teasing stopped – although our relationship then became differently difficult, as he decided we were going to write a book together and pestered me about this constantly.

For people with severe mental disorders, just being around other people is sometimes all that is desired or tolerable. Two middle-aged men at Pine Street Day Centre sat next to each other on a sofa day after day, seldom speaking; if one got up to move around, the other would soon follow. The dayroom on Ward 16 always had a few people, often the same people, sitting there together, smoking silently. 'Company without intimacy' is how the psychiatric literature describes this, and it seems apt. Intimacy is demanding, it pulls hard on our inner resources and threatens our emotional control. Even slight friendships can be taxing, exposing need and vulnerability. For some people this pressure can be too much to handle, at least some of the time. This was quite often true of me, when I was too paralysed by pain to reach out to anyone yet could not stand to be alone. So I would join the smokers in the dayroom, placing myself next to Magda if she were there. Magda would glance over at me and sit on quietly; once she took my hand.

But in better times I was hungry for sociability. In the past Friern had catered quite well to this need. Like nearly all the old asylums, the hospital had prided itself on its recreational facilities, with tearooms, sports clubs, workshops, dances, festivals. Still in the 1970s it had a patient-run radio station and a football team, as well as a successful industrial workshop programme. And even in the run-up to closure, in the late 1980s, there remained a trio of social venues, their names redolent of pastoral ease: the Willow Club, the Orchard Club, the Olive Branch tearoom. Beyond the hospital lay a network of psychiatric facilities – day hospitals, day centres, social clubs – where old friends could meet up and new friendships be formed. By 1990, when I was living in the psychiatric hostel and attending Pine Street Day Centre, I knew this network well – I had plenty of chums inside it and, like the rest of its regulars, I relied on it for daily companionship.

I do not want to romanticize this world. Hanging out with disturbed people can be awful. I know my friends at Friern often found me very poor company. I went several times to the Willow Club (perhaps I was one of the specimens studied by the TAPS sociologists) and found it almost unbearable. Magda, however, felt differently. She knew many people there and enjoyed catching up with them. She was especially pleased one afternoon to encounter a woman who had been a bookseller until she fell prey to chronic alcoholism. To me the woman looked like a mumbling wreck, but Magda chatted with her for a long time while I sulked in a corner. Afterwards Magda scolded me. 'She is a smart, interesting person, Barbara – you need to talk to people before you judge them.' I had been told this before, in very different contexts, but this time the criticism took. After all, I had entered Friern sodden with alcohol – who was I to set myself apart?

Today people with ongoing mental health problems have few dedicated social venues. The day hospitals and day centres have mostly closed, usually in the face of anguished protest. When I ask mental health managers about this, I am repeatedly told that it is not good for service users to spend all their time with other service users, they should mingle with healthy people in the 'community'. It is pointed out to me that mental illness is often episodic; that many people are unwell only intermittently and what they really need is help in utilizing their capabilities during their well times instead of becoming 'career mental patients' consigned to psychiatric ghettoes. There is real force to this argument, and when I repeat it to a service-user activist he strongly endorses it. But it is also a convenient argument, legitimizing yet more swingeing cuts in mental health budgets, and one that leaves untouched the miserable isolation of many mentally ill people in the UK today, sitting alone in their flats with only a television to keep them company. One man who spends his days like this, sitting by himself in front of the telly, told an interviewer: 'It's just like being on a ward again, except there's nobody else there.'[4]

Peter Campbell is a leading figure in the service-user movement. Like many people with long-term mental health problems, he lost most of his 'normal' friends when he became unwell; now his close friends tend to be other service-user activists. 'I've met people who are thought to be extremely strange, extremely frightening, and in fact they are great people . . .'[5] When I met with Peter, I asked him about the situation today. 'The social infrastructure is gone,' he said, 'people are really struggling – but nobody out there seems to care, nobody seems to be listening.'

Communities are not abstractions, they are living entities, products of human relating wherever people find themselves. On my third admission to Friern I became friendly with an Irish alcoholic, a chirpy man who, when sober, was full of ward-improvement plans. 'Games night!' Colin announced one day. 'We'll have a Friday games night; I know lots of great games.' And a few days later: 'A ping-pong championship! Let's set one up – we could run a book on it!' His energy met with little response, including from me, until he hit on the idea of a breakfast club. This was a great success. Every Saturday about a half-dozen of us would sit down and plan the menu. Money would be pooled and Colin would go off to buy groceries. Sunday morning we'd head into the ward kitchen to produce a magnificent fry-up under Colin's bossy supervision. A separate table was laid where we feasted while the rest of the ward looked sourly on. 'It's not our fault that they don't have any,' Colin said, shooing away a woman edging towards his plate, 'I asked everyone if they wanted to join in.' Most of the patients either didn't understand the scheme or lacked the ready cash. Every community has its insiders and outsiders; I was glad to be one of the in-crowd.

Colin was discharged after a few weeks and the breakfast club collapsed. Acute-admissions wards like Ward 16, where people came and went constantly, were difficult places to sustain relationships. Things were different on long-stay wards: here friendships could last for many years. I interviewed Janet Alldred, who was a charge nurse in Friern and one of the people responsible for rehousing the hospital's long-stay patients. 'We made a big effort to keep friends together,' Janet told me. 'Some of the group homes we set up: there are people in them who have been friends for decades.' But there had been a problem. People had been transferred at different rates, with the most socially able going first. 'So the less able people, the ones who had difficulty making friends, they got left behind, which must have been very sad for some of them.'

One of these 'less able' patients was my friend Magda. Ward 16 wasn't meant to contain any long-stayers but in fact it housed several who had been there for many years. 'We're the real vets,' Magda told me. 'They'll never get rid of us.' 'So what will happen when the hospital closes down?' 'We'll hide out, and when everybody else has gone we'll take it over.'

19

Mad Women

I met Magda a few days after my second admission. I was reading in the dorm when a handsome middle-aged woman presented herself to me. She was beautifully turned out in a blue silk kaftan, with a silver-and-turquoise necklace and an armload of silver bracelets. She smiled at me warmly. 'Hi, you don't know me, but I know you.' Who was she? Not a doctor or nurse – maybe a social worker? 'You and Fiona, you were always chatting to each other. I used to lie there' – she gestured to a bed – 'listening to you. I used to do that all the time. I know all about you.' I looked at her, at the bed. Her hair was dark and springy; she had an educated American accent. Was this the blanketed body curled up there, day after day? Magda laughed. 'I know, that's me. But I'm feeling great today.'

Magda is Armenian-American. She grew up in New York and came to London aged twenty. She married a British actor and gave birth to a daughter, now a concert singer. Her gorgeous outfit is because she has just been to hear her daughter singing at the Royal Festival Hall. 'Fantastic, wonderful! I was so proud of her.' By now I was used to hearing about fantasy lives, but Magda looks and sounds convincing. She was a musician herself once, she tells me, a cellist. Well, why not? She was also, I later discover, a scholar. Her field was the visual arts, in particular word-images as religious icons. In 1970 she had published a book on this subject (much-cited, I see now on Google). She had lectured in New York and at the University of

London. She didn't tell me about any of this, but a copy of her book was kept in the nurses' station and a charge nurse showed it to me one day. 'You wrote a book!' I said to Magda, but she was back in bed and didn't respond.

Magda has been in Friern for twenty-one years. Her daughter's birth in 1960 precipitated a catastrophic breakdown from which she briefly recovered, only to go under again. The cycle of illness and recovery recurred with dwindling recovery phases until eventually she ended up on Ward 16, where she is now a fixture. Everybody in Friern seems to know her. She is deeply, irremediably psychotic with the most extreme form of what used to be called manic depression and is now known as bipolar disorder. Depressed, she lies in bed day after day, poleaxed by agony. She is not sleeping; she is lying there torturing herself, rehearsing every bad thing she has done, every bad thought she has had. She tells herself that she is putrid, a lump of shit; she promises herself a future of homelessness and dereliction. Often she prays for death. When I realize the extent of her suffering, I am horrified. I have known excruciating inner pain, but nothing so relentless, so pitiless, as this. 'Why are you still alive?' I blurt out. 'Why aren't you dead?' 'That would be too easy,' she says, with a glint of complacency.

When she is manic, Magda is wild, uncontrollable. She rips her clothes off, screams obscenities, escapes from the hospital and is returned by the police. She is Satan, the Devil, she can do anything. In this state she can be very dangerous. I never saw her like this but Janet Alldred had on many occasions. Janet had known Magda well; at one point in my conversation with Janet she rolls up her sleeve and shows me her arm, with a chunk gouged out of it. 'That was Magda,' she says. When I leave Friern, Magda warns me that if she ever turns up at my door I must not let her in. 'I hurt the people I care for,' she says, meaning it. She claims not to remember her escapades, but later I find out that she does remember them, with a mixture of excoriating shame and lunatic pride. 'I really am Satan, Barbara.'

Janet was fond of Magda. 'Most of the nurses were,' she tells me, 'she could be very loveable, although I don't think she ever knew this about herself.' She could also be unkind, using her fierce intelligence as a weapon of humiliation. 'She was so insightful, she would tell you things about yourself that you didn't want to hear. It was like she was looking straight into your brain.' I'm amused by this description, which sounds so much like my psychoanalyst. Magda had seen an analyst back in New York but it had been hopeless. 'He was a nice man but he sure didn't like me. And I tried to hit him.' This story gives me a small thrill – what would happen to me if I ever tried to hit V? Maybe I too would end up in Friern for ever.

One evening when I was between hospital admissions Magda telephoned me. She spoke at length about being the Devil, about fucking the Devil; she then went on to warn me that 'the Iranians' had poisoned London's water supply. I must promise her I won't drink the water. She was very insistent about this, repeatedly demanding assurances from me that I wouldn't even touch the taps. Then suddenly she broke off the rant to ask about my psychoanalysis: was I finding my sessions helpful? Did V think I was making progress? I laughed at her. 'What kind of crazy are you, Magda?' 'Don't forget about the water,' she said as she rang off.

Two years before Friern closed Magda was put into a little flat in an apartment block with residential nursing staff. I visited her there and found her very happy. She loved the flat, the neighbourhood; she was writing a book about her experiences titled 'Snow in Summer'. A few weeks later I heard that she was back in Friern. She had been found running around the streets one evening, naked, screaming abuse at passers-by and ringing doorbells. I went to see her in Friern but she wouldn't speak to me.

Between her terrible lows and her maniacal highs, Magda could be lively and engaging, as she was on the day that I met her. These times were painfully brief; the morning after we met I found her back in bed again. I left her there, but in the days that followed I tried to coax her out. Finally one morning I yanked the covers off the bed and she rose, eyes rolling. This became a routine. I would arrive at her bedside after breakfast and begin wheedling and nagging, pulling at her covers. If this didn't work, I would start begging her, pleading my misery, my loneliness sitting there by myself in the dayroom. I was shameless, playing on Magda's guilt and shame. 'Oh Christ, Barbara!' Most times she would eventually let herself be dragged up, put on to the toilet, doused in the bath. After I had done this on several mornings, one of the nurses decided to leave her to me. 'Get her up, will you?' this nurse said one morning, 'She has a doctor's appointment.' Magda heard and grinned at me. 'Soon they'll have you giving out the meds.'

Magda was raped in Friern. She was attacked one night in the dayroom, while the night nurse slept – or so she told me. This was certainly possible. Men and women were barely separated at night. The men's dormitory was only metres away from the women's. The nursing station sat between them, with a twenty-four-hour lookout spot. But the station was often empty and after ten o'clock there was only one nurse on duty. He (the night nurses were always men) was supposed to be awake and vigilant but rarely was; one nurse even brought his own blanket to wrap around himself as he nodded off. Male patients sometimes entered the female dormitory. I woke one night to see Ron, a truly terrifying young man built like a sumo wrestler, standing at the dormitory entrance. Eventually he wandered off, leaving me panting with fear. I didn't report this the next day as Magda advised me against it. Ron has knifed people, she tells me, don't get on his bad side.

Peter, a pale slug of a man, patted me on an arm or leg every time he passed me. Finally I kicked out at him and he stopped. One night, when I was sleeping alone in a side room, I woke to see a man in his underpants peering around the door. He leered at me and I yelled and he ran off.

But it was not just the female patients for whom the sexual mixing could be threatening. One evening a young girl was brought in by the police. She was gorgeous – slim and pearly skinned with a waterfall of bronze hair. She arrived distraught and vanished into a side room overnight. The next morning she emerged naked. The nurses rushed to cover her, laughing and joking, and she was sedated. But she would not remain clothed. As soon as the drugs wore off she stripped. She did a turn around the dayroom in the nude, murmuring to herself – an exquisite Ophelia. A few days later I heard that she had been going into the men's dormitory at night and clambering into beds. Rajat was distressed. 'She has to go, it's very upsetting.' Her father came to fetch her and she drifted off on an eddy of trauma.

Soon after I arrived at Friern, Fiona warned me about a nurse who had groped her several times. Fiona was not reliable but the nurse was a nasty little man and I was careful to avoid him. Such things were distressing. But it was the omnipresent threat of physical violence on Ward 16 that really terrified me. 'The most violent ward in north London,' a nurse boasted to me. Patients struck nurses and other patients, tables were overturned, crockery chucked about. An elderly woman had her hip broken by a young lad. And it was not just the patients who were violent. Nurses regularly hurled disruptive men to the floor to inject them. Noisily psychotic men frightened me and I would stand by whenever this happened, silently cheering the nurses on. But these rambunctious men were clearly frightened as well: what effect did this sanctioned brutality have on them?

One afternoon I was sitting in the dayroom when a catatonic female patient was brought in by a nurse. He started tying her into the chair across from me. Ellie may have been catatonic but she was also uncooperative; she kept falling against the ropes. The nurse cursed her. Ellie fell sideways and he punched her. I gasped. The nurse glanced over and winked at me.

Another woman, Polly, who lived in my hostel, was sodomized by an agency nurse. She reported the nurse and the police came round and arrested him. But many incidents went unreported.

Friern had a long history of patient abuse. Most of the old asylums did; like the prisons they had once so strongly resembled, it was just part of their ethos. When David Clark came to Fulbourn Hospital in 1954 he encountered many nurses who felt it their duty to show the patients 'who is boss', often with a kick or a punch to drive the message home. The doctor's duty was not to prevent this violence but to ensure that it 'did not get too far out of hand'.[1] Only the most extreme cases of maltreatment, those resulting in death or serious injury, came to public attention; routine abuse remained an open secret. And the line between legitimate medical procedures and abuse was hazy. Psycho-surgery and electro-convulsive therapy (ECT) were practised throughout the asylum system in the 1950s (ECT is still used today and psycho-surgery is currently undergoing a revival in the USA). I interviewed John Bradley, the medical director of Friern between 1965 and 1976. He told me about a consultant there, a well-known psychiatrist, who was obsessed with uncovering the neurological causes of psychosis, and routinely subjected his patients to lumbar punctures to test their brain fluid. No one stopped this man; no one sued him. 'Incredible, isn't it?' Dr D said when I mentioned this to her. 'To think such things went on . . .'

The silence surrounding these vicious abuses was near-total. (And still is in scholarly circles: I have been unable to find a *single* academic study of staff maltreatment of patients in UK psychiatric hospitals.) But in the 1960s things began to change. As *bien pensant* opinion lined up against the asylum system, the mental hospitals suddenly found themselves the target of unprecedented critical scrutiny. A catalogue of neglect and cruelty was exposed. Many hospitals came in for severe censure, but it was Friern that came under heaviest fire, after Barbara Robb's damning investigation into its treatment of old people with dementia.

Robb, who in the 1960s was well known for her campaigning work on behalf of the elderly, initially came to Friern in 1965 to visit a female patient at the request of the woman's family. The patient was, as her family suspected, being seriously neglected – and she was not alone. Going around the hospital, Robb was horrified to discover hundreds of old people languishing in the back wards, many in states of acute neglect, deprived of their teeth, glasses and hearing aids, and subjected to verbal abuse and physical mistreatment, including beatings, by nursing staff. In 1967 she revealed her findings in a book (*Sans Everything*), which was excerpted in the *Sunday Times* under the headline 'Friern Hospital a dump for geriatrics'. The shockwaves from this exposé, and the public inquiry that followed – which found 'grave problems' at Friern while exonerating it from Robb's charges of gross neglect and cruelty, which in turn led to press accusations of a whitewash – nearly scuppered the hospital. Staff morale was shattered and there was talk of closure. Further scandals followed in the late 1970s when the *Sun* and – more alarmingly for the hospital – the *Daily Telegraph* and *The Times* published reports of patient coercion and neglect at Friern.[2] The *Daily Telegraph* article was based on a leaked internal report into Friern by the regional health authority's own hospital-monitoring team, so its impact was devastating. Friern was said to be 'on the rack', and talk of closure accelerated.

Faced with this crisis, some staff at Friern made strenuous efforts to improve things. But the hospital could not turn itself around. Even under this unprecedented pressure Friern was incapable of self-reform. Jackie Drury was a student nurse at the hospital in 1981. She found elderly women tied down in chairs and beds, without any means of washing themselves except a mobile 'vanity' – a sink and mirror on wheels – with a single hairbrush and one toothbrush to be shared by all the women on the ward. Jackie complained to Friern management about this. But when Dr D arrived as consultant the following year she found the same situation. 'It was disgraceful, no one had done anything,' she said to me. Dr D was also dismayed to find most of the patients on the back wards dressed in ill-fitting clothes from a stack of hospital garments distributed to them at random. 'You could always tell a mental patient by the shortness of his trousers.' But what really floored her was the discovery that some Friern consultants regularly signed off treatment orders without even seeing the patients. On one of her first ward rounds a nurse was surprised when she insisted on seeing the patients before making any changes in their medication. 'Oh no, there's no need for you to do that,' he told her.[3]

The decision to close Friern was publicly announced the following year, in March 1983. Management and staff professed themselves astonished, but if they were, it is difficult to understand why. Jackie Drury later became Director of Islington Mental Health Services, in which capacity, she tells me, she found herself tackling 'performance issues' among former Friern nurses who had moved into the community-based service. 'Some did really well, but others found it hard to adopt a different approach to patients.'

I am not suggesting that most nurses were abusive to patients. The majority I met were civil even under pressure, and some were pleasant in that patronizing way that was once the high point of asylum etiquette. But few were ordinarily kind or friendly. Many were hired in from privately owned nursing agencies. These agency workers were transient, moving from hospital to hospital as needed. Most seemed to know their jobs but had little interest in the patients. Two female agency nurses on my ward spent most of their time chatting with each other, addressing patients only to reprimand them or to bark orders. Male nurses sometimes sparred with male patients but hardly ever chatted with the women. The only nurse of either sex who made any effort to get to know me was Brian, a big Irishman who had been at Friern for many years. (Like all the male nurses, however, Brian felt free to wander into the female dormitory at will, including in the early morning when we were dressing, which made me furious.)

Friern had never been a cheerful place. But the upcoming closure made the atmosphere particularly heavy. Ward 16 was the last acute-admissions ward to be closed in the hospital. As the other admission wards closed, their most difficult patients were shunted over to Ward 16, including several very disturbed women.

Ada was a six-foot-plus Ghanaian woman, muscular and glamorous in her trademark accessory of a huge towel turban. She was notoriously volatile; everyone was afraid of her. Her targets were usually women – Magda had been beaten up by her some years earlier. One morning I went to the dormitory to get a book I had left on my bed. I found it lying on the floor, torn in two. As I picked it up I glanced around and saw Ada standing across from me. Her turban was partly unravelled and she was scowling ferociously. 'You whore! You bitch! Keep away from my things, you animal whore!' The next day she struck me on the arm as I queued for lunch. The nurses hustled her away and Brian called me over. Ada sometimes 'got these ideas' and things could get tricky; maybe I should move wards? I was appalled. Leave my friends? Change my consultant (as I must if I moved wards)? Ward 16 might be hellish but it was a familiar hell, how would I cope on a new ward?

Brian was sympathetic. Instead of moving wards I was put into one of the small seclusion rooms, usually reserved for patients on suicide watch, directly across from the nursing station. But Ada continued to snarl intermittently and my anxiety became unbearable. Finally I bought a pack of cigarettes and went to find her. Her bed was at the end the dormitory. The railing around it was hung with layers of brightly coloured dresses and towels, transforming the bed area into a gaudy tent. No one ever ventured inside this tent uninvited, not even the nurses. I peered through a gap between two towels and saw Ada tidying her locker. 'I have a present for you.' I'm not sure at that moment she knew who I was – her ability to recognize people seemed to switch on and off. At any rate, she waved me in. She took the cigarettes, thanking me prettily. We chatted. I admired her dresses, the beautiful space she had created for herself. She was pleased and asked me about myself. Where did I come from? Did I have any children? No sadly not, I replied. Did she? She looked over at me and smiled knowingly. 'What do you see when you look up at the sky at night?' 'Stars?' I offered. 'Thousands and thousands of stars. Those are mine, my children, all my children. When you look at the stars you are seeing my children.' 'How wonderful!' I said, and it was.

> *I've made friends with Ada.*
>
> *The woman who attacked you?*
>
> *Yes . . . well . . . she didn't really attack me. She just hit me on the arm . . .*
>
> *And tore your book? She sounds pretty dangerous to me.*
>
> *Yes . . . well . . . You know, she's rather beautiful. And she told me about her children in the sky.*
>
> *The sky?*
>
> *The stars . . . she says all the stars are her children . . . Poetic, isn't?*
>
> *Hmm . . . why are you crying?*
>
> *I don't know . . . all those star babies . . . so far away . . .* [Suddenly I can't stop sobbing, which irritates me.]
>
> *Like all those unborn babies that disappeared from your home when you were little? All those children from those unwed mothers?*
>
> [I cannot reply, I am sobbing so hard.]
>
> *Well . . . I can see why it upset you . . . Lost babies . . . up in the stars . . . hmm . . . it* is *poetic.*

'How did you sort out Ada?' Brian asked me a few days later. Ada would now smile at me in passing. I told him I had given her a pack of cigarettes and his eyebrows shot up. Ada's friendliness didn't last long, but she never came after me again.

Thinking back now on this episode, I am struck by the grown-upness with which I tackled Ada. This was not how I usually handled things in Friern. Much of the time I felt and acted like a miserable child. This was not only me – wretched children were everywhere on Ward 16. The old asylums have often been charged with infantilizing their residents. Individuals reliant on these 'total institutions' for all their basic needs – food, housing, medical care, even clothing (I wore my own clothes in Friern but many patients did not) – were said to regress into childish dependency. The claim has force. Ward 16 often felt like a lunatic nursery with naughty – that is, uncooperative or disruptive – patients punished with scoldings or temporary banishment. Temper tantrums were regular occurrences, especially at mealtimes. One man rarely got through a meal without leaping up and chucking his food about. He would be ordered off the ward on to the staircase landing, where he would ring the bell incessantly until someone let him back in.

'Good' – that is, compliant – patients sometimes got little rewards. I was standing at the back of the lunch queue one day when a male nurse called me to the front. Lunch that day was lamb chops – a treat. He piled a plate high with chops and handed it to me. I've no idea why he did this, he had never paid the slightest attention to me before and never did again, but his silly favouritism made me glow. I grabbed the plate and scuttled to a table with dozens of resentful eyes following me. This moment of infantile glee is one of my saddest memories of Friern.

But to blame the old asylums for such babyishness is too simple. Madness is a childish thing; its roots lie deep in infantile sufferings and confusions. I was a baby in adult guise long before I came to Friern. My 'inner child', that much-romanticized self of therapy-talk, was a frantic, furious infant. I had kept this little girl well under wraps, had in fact known nothing about her, until I went into psychoanalysis, where she had erupted with devastating effect. Now that child had the upper hand. My demeanour in Friern was mostly abjectly polite – I saved my rampaging for V – but also extremely anxious and occasionally very demanding. I pestered nurses and doctors for stronger drugs, insisted on being seen by duty psychiatrists. One day I went out and bought a quarter-bottle of vodka and drank it off. This was after several months of no alcohol and I was on heavy medication. Standing in the meds queue at the end of the afternoon, I fainted. I spent that night and all the next day curled up in bed, convinced that I was about to be thrown out. But nothing was said – except by V in my next session.

So — you fainted?

Yes.

I'm not surprised. Do you have any idea how dangerous it can be, combining alcohol with anti-psychotics?

Yes.

You could really have harmed yourself. Or maybe that was the idea? Maybe you imagined if you really damaged yourself I would come to Friern to see you?

No! I wasn't thinking about you! I wasn't thinking about anything!

No? I disagree. I think you were thinking about what to do to make things as bad as possible for yourself — and for me. Not consciously perhaps . . . but I think that was your motivation.

Bad for you?

Yes, of course! You come here and tell me you drank vodka on top of anti-psychotic medication. What if you had died? What kind of analyst would I be then? What would that do to my reputation? There you are, in hospital — that's bad enough! Then you keel over and die. What would people think of me then?

Oh for God's sake!!

So — you don't see any truth in this? Nothing about your self-destructive behaviour has anything at all to do with me, with us, with your hatred of me, with your hatred of the analysis?

[Long silence]

Well . . . ?

[Suddenly I see myself sitting on the toilet, swilling back the vodka. I feel again my deadly glee as the oily liquid scalds my throat.] *OK, OK, OK! . . . I see what you mean!*

Oh?

Yeah. Well . . . I guess I see it . . . I do see it . . . I see what you mean.

Good.

This was the only time I overtly misbehaved, although I squirrelled away surplus meds (in the tummy of a fluffy toy) for the inevitable day when they would be refused me. (I ended up with a small suitcase of pills, which I eventually handed over to a pharmacist, saying they came from 'someone I knew'. 'I hope he or she is still alive!' he said, staring at the mountain of bottles.) Often I was very withdrawn, huddling into myself as I had in early adolescence. But Friern was not my family home and nobody berated me for my sullenness.

Nor, unlike in my family, was I the worst girl in the place. Indeed, compared to Ada and some of the other deeply psychotic women on the ward, I was a paragon. Female lunacy is disreputable. So of course is male lunacy, but not to the same degree. Femininity has always been perceived as having a pathological element, embodied in such familiar figures as the breast-heaving hysteric and the wispy neurasthenic. These figures conform neatly to womanly stereotypes. But the noisy, disinhibited, disruptive madwoman is a perversion of nature, an anti-woman, especially when she is a mother. Salina was a plump Indian in her thirties with a gentle voice and a sweet smile. Salina smiled and smiled until one evening she went berserk and tore her bed apart. Then she tried to hang herself. The nurses dragged her away and Magda told me that Salina had just heard that her daughter was being taken into care.

Friern had always contained more women than men. The asylum was built for this: at its opening in 1851, it had eighteen female wards and fourteen male wards. During the first century-plus of its existence, the average ratio of female patients to male was 1.41:1.[4] This was true of many English asylums, partly for economic reasons (women's financial dependence on men made them likelier to end up in asylums), partly due to women's greater longevity. Were cultural attitudes also a factor? Probably. The division between crazy and sane is gender-biased, with behaviours that are deemed bad in men often labelled as mad in women. Throughout the Victorian period, asylum-keepers complained about the unladylike conduct of their female inmates and made strong efforts to suppress this. Female patients in mid-nineteenth-century Bethlem were put in solitary confinement in the basement 'on account of being violent, mischievous, dirty, and using bad language', while at Colney Hatch between 1865 and 1874 women 'were sedated, given cold baths, and secluded in padded cells up to five times as frequently as male patients'.[5] Female wards were 'noisier' and more 'refractory' than male wards, it was said.[6]

Lunacy in women was, and is, inseparably linked to sexuality. Well into the twentieth century, delinquent sexual behaviour – an illicit affair, an illegitimate child – could get a woman certified for 'moral insanity'. The female reproductive cycle was viewed as inherently pathogenic. 'Women,' an asylum medical officer wrote in 1871, 'become insane during pregnancy, after parturition, during lactation; at the age when the catamenia [menses] first appear and when they disappear . . . The sympathetic connection existing between the brain and the uterus is plainly seen by the most casual observer.'[7] The language sounds archaic now, but the ideas persist: premenstrual tension, antenatal anxiety, postnatal depression, menopausal mood swings – today's woman can sometimes seem almost as uterine-ruled as her Victorian predecessors. I have never been pregnant, but when I was in Friern my periods stopped for a couple of months. When I mentioned this to a duty psychiatrist he asked me whether I intended to have children. 'What? I don't know! Why do you ask?' 'Childless women tend to have poor mental health . . . nature didn't intend women to remain childless.' He smiled at me comfortably: this pitiable woman with her empty womb, her miserable brain.

Nora was a secretary who had worked, so she told me, for ICI. She lay on the floor telling long and lurid tales of corporate shenanigans; sometimes it sounded like she was working up plotlines for a television series. 'You believe me, don't you?' she would ask repeatedly. 'You understand what I'm telling you?' Like Ada, Nora wore a turban – 'to hold my brains in'. Like Salina, she had a little daughter whom she was terrified of losing. Perhaps it was this fear that galvanized her. At any rate, within a few weeks she was bustling in and out of the ward, sorting out a new flat, meeting teachers at her daughter's school. One day she came back to the hospital with her little girl. Together they packed up Nora's things and then departed with kisses and waves, leaving me – and no doubt many of the other women – sick with envy.

Most of the women on Ward 16 had children and many had lost them. This was, and is, common among women with severe mental illness, although how common no one seems to know. Studies of mentally ill mothers are strong on generalizations about maternal incompetence but vague about numbers. And while some profession-als acknowledge the level of personal tragedy involved ('My heart is in chains,' one woman told her case manager. 'The pain never goes away'[8]), others are insouciant about the misery of women facing the prospect of their children being taken from them. Knowing that this terrible anxiety could become a reality 'increases psychiatric treatment compliance' we are told – which is all too easily imagined.[9]

I wept when Nora left. I went along to my analytic session that day and cried inconsolably. My grief was so intense that it astonished me.

I don't know why I'm so upset. I liked her, she was a nice woman, but . . . why I am so upset?
What did you think about her little girl?
Linnie? Linnie was cute. She seemed cheerful, happy. I guess she is really glad to have her mum coming home.
It sounds like it. And what about Nora?
She was so happy! She kept hugging Linnie. She was really affec-tionate. They went off together looking so happy. [More tears]
But Nora's crazy, you said?
Well, yes, I guess so.
But not too crazy to be a loving mother.
No. Not too crazy for that.
And what about your mother?
What?? What about my mother? What about my mother?

Too crazy to be a loving mother? Maybe you would prefer Nora?
That's so fucking silly!
Hmm . . . but true?
[Long silence, then more tears] *Maybe.*

20

Day Patient

I saw a lot of Dr D in the months before I fell into Friern. Once or twice a week I would leave my session with V and go straight to Dr D's office for another half-hour of conversation – my 'top-up therapy' I called this. Inasmuch as I was pleased about anything during this horrible time, I was pleased about this.

It's like being adopted; I've got new parents!
Hmm . . .
You and Dr D – my good parents.
Oh? Is that how it seems to you?
Well, no . . . but . . . well, maybe. [Why is he being so disagreeable? Could he be jealous?]

On the day Dr D admitted me to Friern I asked her if our sessions would continue. No, not while I was in hospital, but they could resume when I left. 'I don't expect you to be there long.' In the meanwhile, I would see her in ward rounds. I had no idea what she meant by this and was too sick and distraught to ask. But I was glad she would be there.

My first stay in Friern lasted a fortnight. A week into this, I was summoned to a ward round. It was held in an area off the dayroom, behind a windowed partition. Sitting beside the partition waiting to be called, I watched shadowy figures moving behind the glass. There seemed to be a lot of them – who were all those people? Coming into the room, I could see Dr D and the charge nurse – but who were all the others? I was introduced. A social worker, a junior psychiatrist, a trainee psychotherapist, some student nurses. They were seated in a half-circle, everyone looking at me. Their faces were neutral but I could smell the curiosity. What had Dr D told them about me? 'Sit down here by me,' Dr D said, patting the chair beside her. Everyone watched as I walked across to her. What sort of performance was I about to give; what would I say, what might I do?

Of all the indignities suffered by asylum patients, the ward round was one of the worst. I still dream about it, this 'ceremonial of control' as R. D. Laing once described it.[1] These days I speak often to groups of people – to classrooms of students, to public audiences large and small. I am a confident speaker, yet even so I often find these occasions demanding. But their rigours are as nothing compared to this command performance, this degrading staging of myself before a roomful of strangers. 'They were pretty horrible, those ward rounds,' I said to one consultant I interviewed. 'Yes!' The fervency of her agreement surprised me. 'So formal, in front of all those people! I'm not surprised some patients freaked out . . . I think I would have.'

Over the next eighteen months, as I went in and out of Friern, I attended many ward rounds. Eventually I learned to focus on Dr D, and to blank out the faces around her. But on this first occasion I had no such stratagems and felt stripped to the bone. My only consolation was that now that I was leaving Friern I would be able to see Dr D one-to-one again. This was a happy thought, and I hugged it to me as Dr D spoke about what came next. How would I feel about attending the day hospital for a while? Yes, sure, that would be fine. [*And our sessions? When will they start up again?*] This settled, she turned brisk, going through my medication and other post-discharge arrangements. [*Why doesn't she say anything about our sessions??*] 'OK, good, that's it then.' She smiled at me and closed the file. I felt myself starting to tremble; I had to ask her.

'Will I have some sessions with you again, now that I'm leaving here? Just us, I mean?' My voice sounded reedy and shrill.

And then – in front of this crowd of anonymous professionals – I discovered my desertion. No more individual sessions. At the day hospital my care would be in the hands of a nurse (a nurse!). I had been abandoned. I couldn't speak. Seeing my face, Dr D said, 'You can ask to see me. I'll be there.' But I never had a one-to-one session with her again.

Dr D won't see me any more. When I go to the day hospital.
What? Not at all?
No. Well, only in ward rounds. Like all the other patients.
Like her other patients? You mean she has *other patients?*
What? What do you mean by that?
The way you talk about her, it sounds like you're her only patient.
Don't say that! That's so stupid! I know perfectly well I'm just
*   another patient!*
Really? Just another patient? Not a very special patient? A unique
*   patient?*
Don't be silly! Of course not!
No . . . ? And if you're not special, she won't look after you any
*   more?*
Won't look after me! No! Yes, yes . . . of course she will! I didn't
*   say she wouldn't.*
No? Are you sure about that?
Why? Why do you say that? Do you think she won't? Do you think
*   she won't??*
Of course she will. It's you who thinks you have to be special for
*   anyone to care about you. I don't think that.*
You don't . . . ?
[Sigh] *No, I don't.*

Friern had no use for strong feelings. Dr D knew all about regression and infantile dependence; my abject neediness was no surprise to her. But what was she to do with such emotions, here inside the hospital machine? She was always kind, and part of me knew that I could trust her even after her 'desertion' of me, but the hospital was systemically unkind. 'I was glad when it closed,' she told me when we met in the summer of 2012. 'I thought it deprived people of much more than it gave them.' She had thought that 'things could be done much better', as she had tried to do them at the day hospital.

Dr D's first job at Friern, in the early 1970s, was at Halliwick House. Like many psychodynamic facilities at the time, Halliwick was run along therapeutic-community lines. Dr D was impressed by the Halliwick therapy groups and the nurses who ran them, especially a nurse named Belle Ridley. Dr D wanted to reproduce the model elsewhere, and a decade later, when she was appointed as one of three consultants at the Whittington Day Hospital, a few miles from Friern, she spotted her opportunity. 'I took it over,' she told me. 'How did you do that?' 'I staged a coup.' The memory still pleases her. She became the day hospital's sole consultant and, working closely with Belle Ridley, who had already introduced some group work into the Whittington, she turned the hospital into a psychodynamic unit based on small-group therapy – which was how I found it when I started there in the summer of 1988.[3]

I arrived at the day hospital feeling like a kid on her first day at school. There were lots of groups – pottery, painting, woodwork, 'relaxation' – and I was ready to sign up for everything. A fresh-faced young man, a patient, showed me around. 'It's nice here,' he said. 'People are friendly.' Afterwards I wandered about, chatting to people in the common room and cruising the bookshelves (bookshelves!) The air was foggy with cigarette smoke; the chairs were plastic, and the carpets were grungy. But compared to Friern, it felt like the Athenaeum. My only complaint was with the hospital hierarchy, or rather the lack of it. It was the nurses, not the doctors, who had primary responsibility for us patients through a care-coordinator system – an innovation initiated by the nursing staff with Dr D's support, which had upset some people during its run-in, and which enraged me now. A *nurse* was going to decide which groups I attended, which drugs I received? Outrageous! I complained to V, who was unsympathetic. 'I'm sure the nurses are perfectly well qualified. You know, for such a radical as you say you are, you're not much of a democrat.'

In fact my own care-coordinator, Josie, was warm and bright and I liked her from the start. My other concern was the group therapy: did I really need to do this? But here the hospital was ahead of me, and Josie told me that I wouldn't be attending a group as a five-day-a-week psychoanalysis was thought to be therapy enough for anyone. At first I was relieved, but after a while I regretted this exclusion, as stories about the groups began to filter back to me. My friend Alex – a one-time pen-pusher in local government who was recovering from a depressive breakdown – adored his group. Arriving back in the common room after his group meetings, he would flop down beside me, burbling with enthusiasm. 'I've never talked to people in this way before! I've never really talked to *anyone* before! It's incredible, it makes me feel fantastic, like I'm a real person!'

Janet, a young secretary who had overdosed on paracetamol and sustained liver damage as a result, took me aside one day and gave me a detailed account of fucking the male nurse who ran her group. Her account of this episode was so deliciously vivid that I was rather sorry it was so obviously fictional. Janet's friend Cary, a gorgeous American with bipolar disorder, danced around the common room with Janet's imaginary paramour in her arms until Janet yelled at her to stop.

This was lively, and for a while I found it easy to join in. I went along to the pottery group and produced a quantity of lumpish bowls and mugs, which I presented to friends (one friend uses hers still). In the 'clerical' group I wrote articles for the hospital bulletin, including one about being in Friern. I painted in art therapy and performed in drama therapy and every Friday afternoon I 'relaxed' for an hour with a roomful of other nail-biters and hair-pullers and floor-drummers.

On sunny days I sometimes went with a few friends to Waterlow Park, just across the road from the hospital. Hanging around there one lunchtime I saw a journalist I knew slightly. She caught my eye, but I ducked into the park café before she could greet me.

This was all OK . . . or OK enough. It was certainly jollier than Friern. But I had no home. I was staying with friends, waiting for my hostel room. Circle 33, the housing association that had taken me on, was opening a new psychiatric hostel in a little house off Caledonian Road. This was where I was going to live – but when? A month passed, then two more months; the house was almost ready, then it wasn't. As delay followed delay I became convinced that the people in charge had changed their minds. Somebody – who? Dr D? Josie? – had told them how crazy I was and they had done a volte-face; I would hear the bad news any day. What would I do? Where would I go? My anxiety levels skyrocketed and my mind darkened with rage: how could they treat me like this, leaving me in limbo? Furious impotence started me boozing again – not much, but enough to add a new worry, that Josie would find out and expel me from the day hospital. Going from Friern to the Whittington, from one refuge to another, I had felt secure for a time. But insecurity was my default position, and now fear rose in me as I became convinced that people were discovering the truth about me – that I was unsalvageable, beyond redemption – and turning away. In a panic one Friday, I asked Josie for stronger medication. On Sunday I had a panic attack and drank some vodka. Monday morning I woke cold with fear.

> *I'm terrified Josie will find out about the booze.*
> *Oh? Why?*
> *I don't know . . . She gave me the meds . . . She always tries to help . . .*
> *And you drank anyway?*
> *Yeah.*
> *You want to show her she can't help you?*

What?

She gives you stronger drugs so you won't drink, but you drink anyway . . . So, so much for her help! Not much use, is she? Is that what you're afraid she'll think? And then she'll hate you for it, for making her feel so inadequate?

But Josie does help me!

Hmm . . . yes, she does. Did your mother help you? When you asked for help, did she help you?

[I'm fourteen. The world has gone blank and cold; I'm in despair. 'I think I'm going crazy, Mum. Nothing feels right to me. I think I'm going crazy.' 'Oh, Barbara! You kids, you know . . . you always exaggerate!'] *Mum said I exaggerated. She said kids always exaggerated.*

When you were breaking down! Does Josie think you're exaggerating, when you tell her how you feel?

No . . . But then why do I behave like this?? Why do I go on drinking, when she's trying to help me?

Because being helped now reminds you of your helplessness with your mother, and that is unbearable. You'd rather turn Josie into an incompetent, and have her punish you for that, than remember what it was really like, when you were a child. Has Josie actually said anything to you about the alcohol?

[Long silence]

Has Josie said anything to you about the alcohol?

[Silence]

Well??

Oh, sorry, I was thinking about something else. Could you say that again?

[Sardonic laughter]

Things eased up a bit after this, and more after my friend Denise moved me into her spare bedroom. I liked living at Denise's. I especially liked living with Denise's baby daughter, Rose. Snuggling Rose, sharing Denise's pleasure in her, I could almost imagine I had a family. This fantasy reopened old wounds, but holding Rose the pain felt less acute than before.

I love being with baby Rose.
Rose?
You know, Denise's baby!
Oh, that baby! And her name is Rose? Really?
Why? Oh, oh . . . you mean like Rose when I was little, the Rose
  who looked after me?
Well, yes! I mean . . . what about her? And didn't you once tell me
  that if you ever had a baby girl, you would name her Rose?
Gosh, that was ages ago! And you remember that?
Of course. I'm your psychoanalyst, remember?

Shortly before Christmas I was taken on a tour of the almost-ready hostel. This awoke a new anxiety. What about my night-time bathing? The boiler was switched off at midnight: would there be enough hot water? I couldn't ask. I lay awake picturing myself heating water on camping stoves, or piling up Thermos flasks of hot water. Control, self-provision . . . who had ever really looked after me, except me? I envisaged the scene in the Circle 33 office when I asked about the water. 'Oh no, we can't do anything about that. We'll cancel your room allocation.' I could see the moues of artificial sympathy, hear the snorts of contempt. Reality-testing this fantasy took all my courage. The support worker at Circle 33 smiled at me across her desk. 'No problem at all, Barbara, the boiler will be left on; is there anything else?' I left her office giddy with joy: I have a home!

The relief swept me into the hostel. I moved in; met my housemates; established a new rhythm (hostel, analysis, day hospital, hostel). The drinking stopped, the drugs went down. I signed up for a photography class at a local college. V and I carried on battling fiercely but sometimes, momentarily, the fog of fear and hate would thin and I would catch a fleeting glimpse of something new, a tiny glimmer of possibility. One day I permitted myself a fantasy: I would get better, and all this horror would lie behind me. Then I would write a book about my bad times. But this idea was so absurd, so obviously and painfully fantastical, that I pushed it away.

Then two months into hostel life, my ex-housemates and I collided again, this time over the value of my house share. Lawyers' letters were exchanged and the air thickened. The clash derailed me, and I sank back into the murderous rage that had almost drowned me the previous summer. Back to Friern. A month there and back to the day hospital. Another fortnight, with more legal wrangling and my possessions disappearing into storage, and back to Friern again. 'You here!' Rajat said when he saw me arrive for the third time. 'Dull out there, eh? More fun in here.'

This time I didn't want to leave. I had my friends, I knew the ropes, I had worked out the quickest routes between Friern Barnet and West Hampstead. And, most important, I felt safe from myself, from my suicidal fury: why should I leave?

Five months into this third admission, at yet another ward round, I told Dr D that I would like to remain in hospital for the time being ('It's a crucial time in the analysis, I think. I would feel safer doing it from here.') She heard me out and then snorted with friendly derision: 'So you want to be an anchorite? No, no, time to move you out.' I was taken with the analogy, but enraged by her refusal to see things my way. 'So, back to the day hospital,' I sighed melodramatically. No, no more day hospital for me; it wasn't working for me any more. I was going to Pine Street Day Centre, over in Finsbury. Finsbury! Miles away! Might as well be another planet! I can't recall what I said then; I don't think I actually begged Dr D to let me return to the day hospital, but I might have. But she was unbudgeable. Pine Street was an interesting place; it would suit me, and I could stay there as long as I needed – which she thought might not be too long now. Really? I probably looked as sceptical as I felt. 'No recovery for me,' I thought. 'I'm stuck in this for good . . . I'm a bloody lifer now . . . you damned doctors can do what you like with me!'

My first glimpse of Pine Street didn't improve my mood. Exmouth Market in Finsbury is a little street tucked into a quiet corner just off Farringdon Road. Today it's a pretty spot crammed with fashionable bars and eateries. But in 1989 it was a dismal backwater lined with betting shops and takeaways. Pine Street Day Centre was on the south side of the Market, in a small building that once had been a 'Maternity and Child Welfare Centre' (the name, inscribed in stone, is still visible above the front door). Back in the 1930s this maternity centre had been part of the Finsbury Health Centre, a pioneering public health facility housed in a magnificent modernist building. In the late 1970s a group of psycho-therapists had been allocated a room in the Health Centre. By then the maternity centre had closed, and in 1980 the therapists took it over and reopened it as a psychotherapeutic day community.

I knew nothing of this history when I arrived at Pine Street. I trudged into the building, a squat brick affair, without even glancing at the inscription over the door. A woman came forward to greet me and show me around. The interior was depressingly dingy: ancient sofas and threadbare rugs, a pool table, a small conservatory with mildewed windows. The toilet was a cubbyhole with crumbling brick-work. I surveyed all this with dismay. Lunch at the day hospital had been served in a modern dining hall; here I saw a wartime kitchen with a vast iron cooker and a few Formica tables. On the counter next to the cooker was a mysterious metal tank, which turned out to be a machine for peeling potatoes. I watched as a sack of potatoes was thrown into the machine's maw. A man flicked a switch and the potatoes hurtled about, emerging a few minutes later pristine and half their original size. The sound was deafening. 'Noisy, isn't it?' I shouted at the man who was watching the process with satisfaction. 'Does the job! Does the job!' The cooking was supposed to be done in rotation, but this man always did the potatoes.

I asked about my schedule. The day hospital had expected me to attend every weekday, but here things were much more informal. About fifty people were registered at the centre but usually only twenty or so showed up. Of these, about a half-dozen had been coming for years; Pine Street was their emotional home. These long-timers lent a certain cosiness to the place. Gladys, a woman in her sixties who had been attending the centre since it opened, did much of the cooking (except the potatoes). When she wasn't in the kitchen Gladys sat at a large round table in the main room, knitting or sewing. Now, on my first day, I sat down next to Gladys and she introduced me to everyone, winking as she pointed out a couple talking – flirting? – in a corner. Her friend Alan, also in his sixties, was accompanied by his pretty spaniel named Blue. 'Does Blue come here often?' I asked. 'All the time . . . he's the craziest of the lot!'

I was introduced to Nigel, Gladys's 'beau' she said, leering sweetly at him. Nigel was in his twenties, round-faced and gentle-mannered. He was quiet most of the time, listening to his voices. 'What do they say to you?' I asked him one day after I'd been at the centre for a few months. 'Nice things mostly . . . I think one of them belongs to my uncle, he died a few years ago. He keeps telling me I should get a car.' Ajala, a delicate Pakistani women in her forties, heard voices too, but hers were vicious. I found her weeping one day in a side room after one of the voices had jeered that she would never see her baby son again (he had been taken into temporary care). Her friend Nadia comforted her. 'They talk a lot of bullshit, those voices of yours.'

The main scheduled activity at Pine Street was psychotherapy, with 'large group' meetings on Mondays and Fridays and 'small group' meetings for everyone twice a week. There was also individual therapy for those who wanted it and were judged likely to benefit from it. Art workshops happened irregularly, but anyone wishing to draw or paint could do this at any time (the artist Bobby Baker, who spent some years at Pine Street in the late 1990s, produced an extraordinary series of 'diary drawings' there, which have been exhibited internationally[3]). A theatre group held regular workshops, and a very successful woodwork class was inaugurated a few months before I left. There were film outings and museum tours and even an occasional sailing excursion, as well as a continual flow of apprentice healers – yogi teachers, holistic-gestalt enthusiasts, guided-fantasy buffs – keen to hone their skills on us. None of these activities (except the group therapy) was compulsory; some people spent most of their time playing pool or sitting around the table smoking and chatting. I went to the large group meetings, played a little pool, did some paintings, stomped and growled my way through a few drama sessions. But I wasn't capable of much. I was – as I had said to Dr D, although I hadn't really believed it at the time – in the eye of the storm, and just surviving took most of my energy.

I can't do this any more.
Do what?
Go on like this. Feel like this. I can't go on like this.
What do you want to do?
I want to feel better, you bastard! I feel like shit, I feel like I'm dying. I wake up every morning and I feel so terrible, my whole body . . . everything . . . And then I come to you, and we talk and I feel even worse . . . and then I go on to that horrible dreary place . . . and I would rather be dead. I should be dead. I feel like I'm dead.
Is it horrible and dreary?
What??
Pine Street. You just said it's horrible and dreary.
Oh. No . . . it's OK. Dingy, you know . . . but it's OK.

Unlike the day hospital, where the nurses were responsible for the group therapy, Pine Street employed three psychotherapists: Ben and Louise, both of whom had come there from the Henderson Hospital, and Nick, the director, a group therapist with long experience of therapeutic communities. Nick was a therapeutic species that was new to me – good-looking, sardonic, compelling. He wore his authority lightly, with a facetious air, and rode a big motorbike. Had I known more about the therapeutic-community scene, I might have recognized him as a familiar type: the (white, male) leader who dominates through force of personality, and whose influence stamps itself all over the community. Such charismatic men have featured throughout the history of the therapeutic-community movement, from pioneers like Maxwell Jones and Tom Main through to the anti-psychiatry communes of the 1960s and 1970s with their counter-cultural gurus such as Ronald Laing and David Cooper. In the late 1970s, Pine Street's predecessor, the Paddington Day Hospital in west London, one of Britain's first non-residential therapeutic communities, imploded amid internal conflict and public scandal, mostly focused on its director, Julian Goodburn, a *Marxisant* psychoanalyst with a deliberately confrontational style who inspired fierce loyalty in some PDH patients and bitter antagonism in others. Like most radical therapeutic communities, PDH claimed to be democratic, but Goodburn's know-it-all hubris and psychoanalytic purism (he refused to sign patients' medical certificates on the grounds that this would be anti-therapeutic) led to charges of autocracy from patients and management. A decade earlier a coalition of staff and patients had defeated an attempt to close the PDH (this struggle, which became a cause célèbre on the left, was the seedbed for the service-users movement). But this unity had soon collapsed, and in 1976 patients' complaints had prompted two official inquiries, which resulted in Goodburn's dismissal and the PDH's closure three years later.

Nick was not a leader in Goodburn's mould. He was febrile and sharp-tongued and liked to play favourites, but he was no left-wing guru, and Pine Street was no flagship for libertarian democracy. Perhaps it had started out with radical ambitions, but late-1980s Britain was a very different world from the one that had produced psycho-politicos like Laing and Goodburn. Nevertheless, thorny issues of power and leadership hovered over Pine Street. Group therapy, as promoted by the Institute of Group Analysis, of which Nick was a leading light, was meant to be 'group-centred and not leader-centred', 'a form of psychotherapy by the group, for the group, including its conductor'. But it would have taken a brave person to interpret Nick's jibes, or to challenge his Olympian pronouncements on group interactions. Stories of such challenges were part of Pine Street folklore but I never witnessed any, although I occasionally heard mutterings from people smarting from Nick's barbs. But he was mostly admired and liked, and watching him in group meetings I saw that he could be very effective. Even the most difficult or withdrawn people often responded positively to his tough-love ministrations. But not me – he reminded me too much of left-wing ayatollahs I had known, and anyway he didn't pay much attention to me. In fact all three therapists mostly left me alone, which annoyed me occasionally but overall suited me well. I was deep inside the beast's belly now; every day was a struggle, and what I needed most was solitude with other people, which Pine Street gave to me. 'You seem very withdrawn,' Louise said to me one day in January 1990. 'Not withdrawn,' I replied. 'Just staying alive.'

21

The Hostel

It's pretty good, you know.

What?

Pine Street. I mean, Nick can be pretty annoying but I like Louise and Ben . . . I know I'm not exactly a typical patient . . .

What is a 'typical patient'?

You know what I mean!

Hmm . . . I suppose, but I don't much like that 'typical'. In my experience there's no such thing as a 'typical' patient . . .

[Loud sigh] *Anyway, I think it's good. People working together . . . I like that.*

Yes, working together, cooperating . . . it is good, isn't it? A good way for people to be helped.

Hmm . . . is that a hint?

What do you think?

Pine Street provided community care of a kind that is nearly extinct today. 'Community psychiatry' used to be a synonym for places like Pine Street and the Whittington Day Hospital, institutions where staff and patients collaborated to heal madness, or at least to mitigate its sufferings. Pine Street was ramshackle and idiosyncratic but it was also – director Nick's foibles notwithstanding – humane and effective. Without it, I would almost certainly have landed back in Friern for a fourth time. I saw people at Pine Street who were locked into chronic illness, but I saw others who were changing, gaining strength and moving forward, sometimes quickly, sometimes slowly. Nobody was rushed; those who couldn't or wouldn't 'get better' were treated kindly. Mental health officials itched to close the place down. 'People just stagnated there,' I was told.

My hostel offered another variety of community care, the kind once labelled 'aftercare'. Psychiatric aftercare for Friern patients had a long history. In 1879 the chaplain of Colney Hatch, Revd Henry Hawkins, organized a meeting of psychiatric notables to deal with what he called the 'unhappy situation' of women discharged from the asylum with nowhere to go. The meeting resulted in the founding of the Aftercare Association for Poor and Friendless Female Convalescents on Leaving Asylums for the Insane, which established a number of 'convalescent homes' for such women. Today Revd Hawkins's association is known as Together and still runs psychiatric hostels. Other mental health charities do likewise, as do not-for-profit housing associations like Circle 33. In 1989, when I moved into my Circle 33 hostel, this sort of 'voluntary sector' psychiatric accommodation was thin on the ground. Now it has much increased, but the really big increase has been in commercial provision, which has grown by leaps and bounds since the asylum closures.[1] (The health care market intelligence agency Laing & Buisson issues regular updates on the money to be made from psychiatric residential care.) If I were looking for a hostel today, I would have a dizzying range from which to choose, offering me everything from 'enhanced life chances' to 'increased self-knowledge and self-worth' – if I could afford such delectables, that is, since only people who have been legally detained have any entitlement to public funding for supported housing. Everyone else is means-tested, and the fees are very high.

'Community care' has of course always existed – in people's own homes. Even at the peak of the Asylum Age, most people with mental disorders were cared for by their families. Revd Hawkins's residences were for 'friendless' women. I was not friendless but I needed more care than my friends could give, and my lack of family in the UK was probably decisive in securing me a funded hostel place. It didn't occur to me to be grateful for this: I took public health care for granted. But nor did I expect enhanced life-chances or elevated self-worth. I expected basic, no-frills accommodation with some practical and psychological support – and that is what I got. At the time, it made all the difference.

I was an old hand at group living, having spent nearly two decades in shared houses. I had never enjoyed it: I was too insensitive to the needs and feelings of others, and too anxious about other people's feelings towards me, to be a happy cohabitee. I thought I was sociable but in fact what I really liked was being on my own with others near by. I needed people around me, but took little pleasure in their company. Housemates impinged too much on me; I was dependent on their presence, but their proximity made me anxious and miserable. I had managed this conflict throughout my twenties but by the time I went into psychoanalysis it was becoming unbearable.

My last shared household, the one I was asked to leave, had been set up in 1982, when my then-partner and I had bought a big house in Highbury with another couple. He and I split up soon after. He left, and I stayed on with the couple. They had small children, demanding jobs, busy lives; soon they also had a madwoman in their attic. I hid away in my top-floor rooms with my vodka bottle, or slouched around the house looking frail and wretched. As things worsened, and I struggled to control my drinking, I insisted that they lock up their alcohol in a cupboard that I bought for the purpose; amazingly, they complied. Sometimes I refused to do my share of household chores. Things weren't always as bad as this – I was fond of their little boy, and he and I became good chums. But the situation was impossible. Part of me knew this, but I could not act to change things; I left that to them – an additional burden. My half-acknowledged awareness of what I was putting them through worked on me constantly. I yearned to go long before I was pushed out – but where would I escape to, how would I manage?

So, when I finally found myself in the hostel, my first reaction was elation. Walking into the house, I was swept by a sense of freedom so powerful that I wobbled and had to sit down. I had sloughed off an impossible emotional burden. Here there was no one to be well for; no need for subterfuge; no fear of homelessness if I couldn't perform the rites of friendship. I could be as I was, without terror of rejection. Here, in a home for 'vulnerable adults', I had my first glimpse of adult independence. The thrill didn't last long, but the liberation was real. I could want and need other people without enslaving myself to them (and then hating them for it): a lasting revelation.

The hostel was a three-storey Edwardian terrace in a treeless road surrounded by council housing. There were five single bedrooms, two bathrooms, a kitchen with individual lockers for food supplies, a small lounge. The furnishings were IKEA-basic. The bedrooms were tiny, with space only for some clothing and a few books, so friends gathered up my possessions and distributed them among a dozen households for safe keeping. I didn't think I would ever have anywhere to put these things again. When they were returned to me two and a half years later, after I had moved into my new flat, not a single item was missing.

My new housemates were all women in their thirties and forties. The room next to mine was occupied by Mary, a rectangular Mancunian with tattooed arms and a confident manner. She was friendly to me but I glimpsed aggression. A few weeks after I arrived she attacked another woman and was immediately evicted. Her room was taken by Shana, a Jamaican girl who wore no shoes and spoke in a whisper. Shana was an eerie presence. She would glide silently into the kitchen and cook up spicy-smelling stews that she ate in her bedroom. The bedroom walls were thin and at night I would sometimes hear her call out, little hiccupy cries that I didn't investigate. One night the

Devil tried to clamber into bed with Shana and she threw herself out of the window. Her back was badly injured. She returned briefly to the hostel from hospital, but by now her lunacy was so acute that the rest of us protested and she was removed.

Betty was the oldest in our group, a slovenly, malicious woman in her forties whose only problem, it seemed to me, was her profound stupidity. Betty was the woman Mary had attacked, and while Mary's violence terrified me I sympathized with her attitude to Betty, whose sottish spitefulness made me want to smack her too. Betty slobbed around the house, eating whatever she could lay her hands on; or she lay in bed listening to deafening pop music, ignoring all requests to lower the volume. She was flabby in peculiar places, and moved her limbs as if she were in dispute with them, constantly banging and bruising herself and then wailing like a small child. But there was a mystery about Betty: she had a boyfriend, a handsome, well-spoken Nigerian who turned up most weekends to take her out. He brought her gifts, sometimes items of clothing; he was courtly, charming. 'What on earth does he see in her?' I said to Polly, my friend in the house. One day he brought Betty a kitten. She was thrilled but the rest of us feared for the kitten. Soon Polly and I found ourselves taking care of the kitten, after he scratched Betty and she kicked him across the kitchen.

Polly was the youngest hostel resident. She was middle class, well educated, and suicidally anorexic. Her anorexia ruled her life; everything connected to food was torture to her. I had known other women with anorexia, but never anything like this. Polly had been hospitalized many times, subjected to force-feeding and a host of other regimens designed to keep her weight at a survivable level. When she moved into the hostel she was skeletally skinny but not actually starving; after a while she even put on a few pounds – but this was intolerable and she soon stripped them off. She exercised ferociously, lying on her back bicycling into the air for at least an hour each day.

At first I was delighted to have Polly around. She read a lot and we chatted about books; we went for walks and even to an occasional film. She had been in therapy, and knew something about the causes of her disorder. I listened when she talked about her parents' miserable marriage, her younger sister who was also anorexic. I didn't comment on her meals, which over time became increasingly frugal, or the elaborate rituals of weighing and measuring that accompanied them.

Like me, Polly was hospitalized twice while she lived in the hostel. As I went back and forth to Friern, she went off to an eating-disorders ward in a general hospital. On her second admission I went to visit her. I expected somewhere like Ward 16; instead I found myself in a bright, homey place filled with overstuffed furniture and fluffy toys, the walls lined with pictures of pop stars and holiday beaches. Young women wandered about, skinny arms poking out of oversize T-shirts, ears plugged into Walkman stereos. Polly showed me the dining room where they assembled five times daily for meals and snacks, watched over carefully by the nursing staff. 'I'm doing well,' she said to me, but she looked cadaverous and I smelled vomit on her breath.

My feelings for Polly soured as her condition worsened. Back in the hostel, with her weight still dangerously low, her meal rituals became increasingly intricate and I often heard her vomiting after she ate. This was depressing enough, but then she began regaling me with terrible stories about women who, after years of anorexia, embarked on huge eating binges until they became horribly ill. 'Sometimes their stomachs burst.' I could feel her watching my reaction. 'Do you go on eating binges?' I asked her carefully. 'Oh no,' she said, 'Well, not big ones – maybe a packet of biscuits or something like that.'

One night I heard Polly running up and down the stairs. The noise woke me and I was annoyed. 'What the hell are you doing?' 'I have a leg cramp,' she said. 'I was just easing it out.' A few days later she called out to me as I was passing her bedroom. She sounded panicked.

I opened her door and reeled back. The bed mattress was on the floor, covered in vomit; Polly was lying across it, shaking violently. I kneeled down next to her.

'Polly, what's happened to you? What's going on? What's happened to you?'

She looked at me with an expression I couldn't read. 'I've eaten rather a lot.'

'What? What have you eaten?'

'A loaf of bread.'

'A whole loaf?'

'Yes, a big loaf. And some tins of cat food.'

'Cat food! What??' Suddenly I was so enraged I could barely speak. 'You've eaten, eaten . . . cat food?? What the hell? What are you trying to do, what are you trying to do?'

I yanked Polly up by the shoulders. 'What are you doing? Why did you do this? Are you trying to kill yourself?'

Now I saw the slyness in her look, the little moue of satisfaction at my distress. 'Oh, Barbara, I'm sorry, I'm sorry.'

'Oh for Christ's sake!' Her shaking was terrible, her lips were blue. 'Oh Christ!' I ran downstairs and got Betty – 'Help me to get Polly into the car, I have to get her to hospital.'

Polly nearly died that day. Soon after I got her to the hospital she began to convulse. Starvation had short-circuited her brain. She was raced on to a ward and hooked up to machines and drips – where I found her a few days later, when I could bear to go and see her. She looked terrible, black bruising around her eyes and sheet-pale. But this time when she looked at me I saw no complacence, no sadistic pleasure in what she had put us both through, only a sad wistfulness that made me blink. 'Sorry, Barb.'

Polly and I shared our Circle 33 support worker, a trainee psychotherapist named Florence. Florence was a honey: kind and sweet-mannered with a tough shrewdness that I admired. I wanted to *be* Florence, I decided; she was my role model. When she left for a new job, towards the end of my time in the hostel, I wept bitterly, and was sulky with the woman who replaced her.

Florence came to see me a few days after Polly was hospitalized. She thanked me for looking after Polly. 'You seem to be getting stronger, Barbara. You dealt so well with Polly, with the hospital and so on —' I was startled: was I feeling stronger? I eyed Florence warily. Might this be a preliminary to chucking me out? Florence saw my expression and moved on hastily. 'I know life is still really tough for you. I just wanted to say thanks.' She paused and looked hard at me. 'You know, it's really good having you here, good for the others, I mean. I know Polly feels that — and so do I.' Now I was dumbfounded. Good for the others? Florence was smiling at me. I began to cry. Then, to my utter astonishment, I reached across and took her hand. 'Thank you, Florence,' I said. 'And you're very good for me. Thank you.'

In my session the next morning I tell V about Florence's visit. I'm embarrassed when I get to the bit about taking her hand and thanking her, but then the feelings of gratitude rise up again and I feel warm and happy.

> *I like Florence so much. She's smart, and kind. I feel like I can really talk to her.*
>
> *Unlike me, you mean?*
>
> *You? Who's talking about you? I am talking about Florence. She's such a good listener. She will make a good psychotherapist, I think.*
>
> *Unlike me, you mean? Sorry, you've already said, this isn't about me.*
>
> *No, it isn't! Why are you teasing me?*
>
> *I'm pleased with you.*
>
> *Are you? Pleased with me? Why? Because of Polly?*
>
> *Because of Polly, but also because you are learning to use what's made available to you. To find it helpful instead of just attacking it.*

Florence?
Yes, and the hostel.
You?
[Laughing] *Well, let's not get carried away!*

PART THREE

22

Change

There is nothing . . . that can solve the problem that other people actually exist, and we are utterly dependent on them as actually existing, separate other people . . . [E]verything else . . . follows on from this.

Adam Phillips, *Missing Out*[1]

We become who we are through relationships. The 'I' is born at the interface between self and other, the helpless and the help-giver, infant and parent. As babies we learn about ourselves via the minds of those around us; inchoate sensations take on shape and meaning through the responses of others. Selfhood surfaces on a tide of recognition: this is who you are/this is who I am. We human beings are dependent creatures who discover ourselves in communication with others, spoken and unspoken, conscious and unconscious. Without such communication the individual remains undiscovered, lost in a limbo of unintelligible being.

30 November 1989. I arrive at my session fit to be tied, as the saying goes. I lower myself on to the couch and go very still. Then I hurl into action. I throw myself at V, punching him in the chest, biting his face, kicking his balls, twisting his arms until they snap. Finally, in an ecstasy of fury, I devour him wholesale, beginning with his head. Afterwards I lie satiated, shaking, listening to him breathing quietly

behind me. The tumult subsides, my heart rate slows; another kind of image appears in my mind's eye.

I am an outline.

What do you mean?

What I said! . . . An outline . . . I mean, just a line around nothing . . . A line and nothing else . . . no me . . . there's nothing there, just a line. A line . . . [Long pause] *A dirty line, a greasy, filthy, dirty line . . .*

Like your dirty face?

Yes! A line around nothing . . . well . . . with bits and pieces in the middle of the nothing . . . lots of crap scattered around . . . dreams maybe . . . feelings maybe . . .

Feelings?

Yes! Yes . . . All these horrible feelings scattered around . . . nothing holding them together. Just an outline . . . I'm just an outline. [I lift my hand into the air and sketch a wavy line.] *See? Like that . . . see?*

Yes, I see. [Pause] *Where have your feelings gone?*

What? What?

You came here very angry. What has happened to those feelings?

I don't know. I don't feel angry now.

How do you feel? [His voice is unusually gentle.]

I feel OK . . . [Long pause] *Yes, I do! I feel OK really . . . I feel OK here . . .* [Tapping myself on the chest]

Good.

This outline image, once it had appeared, recurred constantly. I saw it in my sessions and in my dreams; I even tried drawing it in my journal. Its horribleness is hard to convey, but it seemed to encapsulate the combination of acute tension and nebulous despair that was my usual state at this time, except when I filled with pain or exploded with rage. This outline demarcated the walls of my personal hell, my doorless prison – nothing out, nothing in. Nothing anywhere that made sense of anything, no connections, no history . . . only the amorphous, timeless misery of a mind alone.

Except that I was not alone. For nearly eight years I had spent fifty minutes a day, five days a week, with V. During that time I had experienced him mostly as an adjunct of me, another bit of flotsam drifting about in my inner world. But the reality of his separateness, of his implacable otherness from me, pressed on me constantly. I fought it off ferociously, like someone about to be violated or eaten. All evidence of his actuality was obliterated, pulverized – only to reappear in the next session as he brazenly ignored my no-go injunctions. It was unbelievable, intolerable . . . and also, finally, after interminable bitter battles that left us both exhausted, incredibly exciting. It was the arrival of this excitement, in the spring of 1990, that signalled the change in me. But the months preceding it were a maelstrom of pain and confusion that nearly capsized me.

The confusion was important. I had come into psychoanalysis full of fixed convictions about who I was and what mattered to me. These had nearly killed me. Now I understood nothing. My journal entries became a litany of questions. I dreamed about lost letters, scrambled messages, failed telephone calls. At times the muddle was so overwhelming that I nearly blacked out. At other times it felt oddly comforting, as if perplexity were my natural state and everything else was forced and artificial – but this was rare respite. I began having terrible panic attacks. These would start deep in my body and then spread through me, closing my throat and stiffening my muscles. By the late winter I was having these attacks frequently, especially on weekends when I often spent entire days in bed, immobilized by fear. My drinking, which I had kept down since leaving Friern, increased again, as did my drug-taking from the large store of pills I had squirrelled away while in hospital. I began to think about returning to Friern – how much longer could I go on like this?

Yet from the depths of this misery sparks of optimism began to flare – little cheer-up messages relayed from somewhere I couldn't perceive. 'I am going to be well,' I write in my journal one day. 'The second half of my life is going to feel utterly different from the first.' How did I know this? My symptoms were changing. The deathly exhaustion had vanished, swept away on flood tides of rage; some compulsive tics had disappeared. Daily life was horrible but I was managing it better and – very important – learning to use the help I received. Pine Street and the hostel both felt like secure places to hunker down, and I valued them for it. My friend Anna, who lived in a collective house in south London, had opened her home to me and I went there gratefully.

Anna's generosity was proverbial. For nearly a year I took full advantage of it, turning up at her door almost every weekday after Pine Street closed and staying with her most weekends. I still couldn't manage without a caretaker, but the persecutory pressure behind this need had relaxed somewhat. Anna had a teaching commitment in the United States every autumn. She would be gone for two months: how would I cope? A few years earlier I might have launched a full-scale campaign to inveigle her into staying with me. Now, after some anxious strategizing, I contacted Islington Social Services, who told me about their 'adult adoption' programme. I could have a room in the home of a retired nurse who would provide meals and generally be there for me: would this suit me? I went along to meet Shirley the nurse, an Irishwoman in her sixties with a nice smile and a cosy voice. I signed up to start with Shirley in September, when Anna left for the States, and heaved a sigh of relief.

Looking back now, I can see that these developments signalled the change that was coming. At the time, however, they seemed trivial compared to my suffering. In early 1990 a close friend and one-time lover with bipolar disorder killed himself. He had been in psychoanalysis and it had taken him into places I knew all too well. 'Hang on,' I begged him when we met up. But he had suffered much longer and harder than me. One evening he checked himself into a hotel and died there. The day I got this news I saw a car coming towards me and watched myself going under its tyres. The car passed and I was surprised to find myself still on the pavement. I had never been entitled to life. Now death was coming at me; it was just a motion away.

I slept, when I slept, clutching the pebble V had given to me. In my sessions, V and I talked about how Dad had teased me sexually – all that systematic stroking and caressing – and how Mum had connived at this. The pain that followed these sessions was unbearable, but if I drank to ease it I became panicky. 'I think I must go to hospital,' I told my journal. 'I am going to die like this. I can't go on like this.'

Yet still hope flickered. 'I feel hope and don't dare to feel hope,' I wrote. 'Too bad to hope?' Feelings I had long hidden from V – extremes of hate and envy but also passionate sexual desire – erupted in my sessions and to my surprise this did not embarrass me. I saw myself fucking him, sucking his penis, having his baby. 'Just one?' he said about the baby fantasy. 'No, no – lots! Thousands of babies!' I said, remembering the mad Ada. After years of standing my ground, defying all V's attempts to move me, I was now careering around inside feelings and fantasies that were mostly viciously destructive, but also occasionally, to my astonishment, deeply affectionate. 'I do not know who writes here,' I confided to my journal. 'Who is she? I don't know her any more.' And then, a line further down: 'Who do I think I am talking to here?'

One weekend I drank myself into oblivion at the home of some close friends, Lori and Geoff. I had gone there for dinner; they had cooked a nice meal and afterwards I listened as they talked about their home-renovation plans. I stayed overnight. When I didn't appear the next morning Lori came into the bedroom and found me semi-conscious, my duvet splashed with vomit. She was frightened and furious: what was going on? Where had the booze come from? I had told them I didn't have any with me. I was angered by her anger: didn't she know the hell I was in, the exorbitance of my suffering? But I was overwhelmed with guilt too. I couldn't stand the shame of taking responsibility for my actions.

I can't stand the shame of my responsibility for what I do.

Hmm . . . yes. But maybe not just that?

No? What do you mean? My God, it's so bad. I feel so ashamed.

Of what?

I just said — I don't take responsibility for what I do, for how it affects people.

And how does your behaviour affect your friends?

You know! It frightens them, it upsets them. It makes them very angry with me.

Then why do you do it?

What?

Why do you do these things? Go to people's homes and get drunk? Why did you do that to Lori and Geoff?

What? I wasn't trying to do anything to Lori and Geoff! I just didn't think about them at all . . . I felt so terrible, I couldn't stand the pain. It had nothing to do with them! Now you're getting angry with me!

I'm not angry with you. I asked you a question. Why do you do these things that disturb your friends so much?

I'm not thinking about them when I do these things! It's not about them!

No? What's it about?

It's about the way I feel, how terrible I feel. It's got nothing to do with them!

No?

No!

I am so furious I cannot speak. I lie rigid. The room is quiet except for our breathing. After a long while I feel my fists unclench. My heart begins to slow; my mind ticks over cautiously. Lori and Geoff: their new home, the pretty garden they were tending when I arrived. The nice meal, their friendliness . . . the sounds I had heard from their bedroom as I lay awake. And me: my home in a psychiatric hostel;

my booze; my empty bed, my empty days. I am soon going to turn forty.

I grunt experimentally and hear him sigh behind me.

I am so miserable. I want to spread the pain around.

Yes. But then that makes you feel worse.

Yes. Yes, I see. I think I see. I see. [My heart is slow, pumping tears . . . everything is loose. I start to cry. I cry until the pillow beneath my ears is sodden.]

It's time.

The ice in my muscles has melted; I wobble when I stand up. As I leave the room I turn to smile at him. He looks at me, surprised, then smiles back.

The next morning I wake on a surge of pain so strong I cannot breathe. I swig vodka and lurch to my session. As I walk over to the couch I glance at V. He looks tired: what's the matter with him? Anxiety sluices through me.

I feel horrible, really horrible.

Hmm.

I don't know what's the matter with me. I felt OK after the session yesterday.

[Silence]

You were right, you know.

Really? [His tone is sarcastic.]

Yesterday, about Lori and Geoff, why I did what I did. Like I said, I'm so miserable, I want to spread the pain around.

Hmm . . . yes.

Yes? You don't sound very certain.

No?

No? Yes? What is it? What do you think? God, I feel awful.

[Silence]

What? What do you think?

I don't think you want to hear what I think.

I just asked you, didn't I?

That's true.

So??

[Silence. Is he still breathing? Has he gone to sleep?]

Why don't you say something?!

So that you can attack it?

Oh for Christ's sake! I said I got drunk at Lori's because I am so unhappy, I want to make other people unhappy too.

Is that what you're doing here today?

What? What are you talking about? I told you you were right yesterday.

And now you attack everything I say.

I'm not attacking you!

[Silence]

[I shuffle around on the couch, stare out the window at the early spring light.] *What were you going to say? I do want to hear it. I feel so miserable.*

Hmm . . . Why do you think you are so miserable?

Why? Why? Christ!

I know it sounds like a silly question.

Yes, it does!

You say you want to make other people unhappy because you're so unhappy. You say you come here drunk because you're so unhappy.

I'm not drunk!

No? Why are you so angry and miserable today?

Why are you so fucking grouchy?! Are you feeling tired?

Why would I be tired? Did we do too much good work yesterday? What??

You heard me. You told me I was right yesterday, now you attack me. You tell me I'm tired: have you tired me? Have you worn me out?

I don't know what you're talking about.

Envy. I'm talking about envy.

Of who? Envy of who?

Of whom!! [Sigh] *Of me! You envy me for being able to think
straight, for having wit enough to understand you. You envy me
for having my own mind, my own life. You envy Lori and Geoff,
for having their own lives.*

Oh.

*It's the envy that's killing you. You don't vomit all over Lori's bed
because you're an unhappy person. It's envy that you're spewing
up — over her and Geoff, over me — that's what is devastating
you; that is what's making your life such a living hell.*

[Silence. The nausea is rising, I can feel it in my throat.] *You've
said these things to me before.*

*Yes, many times, and I'll go on saying them until you hear
them.*

About envy? About my wish to destroy people?

*To destroy their well-being, yes, and their ability to think about you
in ways that you can't control. And your guilt about that.*

[Am I going to throw up? I feel incredibly sick. I feel . . .]
I thought you looked tired when I came into the room.

Too tired to think about you any more?

Maybe.

What about Lori? Do you think she'll think about you any more?

She rang me today to see how I am.

Ah. How did she sound?

Fine. Concerned.

About you?

Yes. [I seem to have been crying for a long time without real-
izing it. I seem to be flowing away on a river of tears.]

Well.

[I could lie like this for ever, floating on my tears.]

It's time.

There are no miracles in psychoanalysis. I had had literally hundreds of sessions like this one. I knew the drill, I knew the lingo; I could cite the classic psychoanalytic texts on envy and guilt, chapter and verse. Nothing had changed, nothing had made any difference. I was going to die like this.

23

Separation

In February 1990 I stopped drinking. I had begun to feel sick after I drank. I couldn't tell whether it was my body or my mind that was saying no to the booze. But I had had enough. 'Whatever is going to happen has got to happen,' I instruct myself sternly. 'I need more courage.'

19 March 1990. I am sitting in the waiting area before my session, buzzing with eager anticipation. V comes out of his room on to the stair landing and looks down to where I am sitting. 'Come up, please.' I go up, enter, lie down, we talk. After a time his words start to grate on me; my excitement darkens. Why was I looking forward to seeing him? Everything he says is inane, beside the point . . . he's a blundering idiot. I feel myself going rigid with rage . . . I want to kill him, to tear him up. I hear my voice go tight and shrill; I think I am going to vomit with furious disappointment.

Why?? Suddenly the question rises in me like an interrogative belch. My mind screeches to a halt. I lie without speaking for a very long time. I hear cars passing by, V breathing quietly behind me. The tree across the street flaunts its springtime leaves, tender baby greenery. Gradually my body loosens; I feel the tears coming.

So strange, all this is.
Yes.
You know what I mean?

Yes, I think so. Changing is very strange.

Change . . . Am I changing? Yes . . . You know, I think maybe
I am.

Hmm.

You know, I think maybe you are saving me. I think you are saving
my life.

Hmm . . . well. We . . . We are saving your life.

Yes, we . . .

It's time.

The session after this one was horrible, so pain-wracked and desperate that I almost bought a bottle of vodka on the way home. But
by the evening I was calmer. A new thought occurred to me. 'The
truth is, I don't know what tomorrow's session will be like, and this
is what keeps me going.' Having written this sentence I sat and looked
at it. Well, well. My omniscience had slipped away while I wasn't
looking. Like a bad dream that disperses on waking, my deadly
power – over V, over everyone and everything in my lonely universe –
was evaporating.

But not without a last-ditch fight. My power has been my lifeline;
how can I let it go?

I dream that Mum is my psychoanalyst: 'I cannot hate my mother.'
I dream about a man who is involved with me in some indeterminate
plot. He and I run along a street together. Two kittens rush up to us
followed by savage dogs. I smash one of the dogs on the head until its
eyes bug out. 'I'll kill you if it will save these kittens!' I scream at the
dog. But this is very bad: 'It is the most natural thing in the world for
a dog to kill kittens,' a voice scolds me. I wake with my upper body
twisted in pain. 'Hate-pain,' V tells me. 'And who are those kittens?'

But I won't reply. I won't play any more; I have had it with V and
his psycho-shit.

I dream I am curled up with V in a field full of barbed wire. 'It's a
battlefield,' a voice says.

I dream I am lying on V's lap and he is stroking my back as he makes interpretations.

I dream of Mum describing her childhood to me; as she speaks she bites her nails, tears her hair, vomits. 'I used to vomit too,' I tell her.

I tell V only the last of these dreams. I am furious with him, so angry I can barely speak.

Such a horrible, nasty dream!
Yes.
Yes, yes, yes! Well then! What does it mean?
It means you are separating from your mother.
Hah! You've been saying that for years . . . For years! And here I
*   still am, stuck here, stuck with you.*
[Silence]
Well, you have, haven't you?
[Silence.]
Why don't you say something?
So you can attack me?
Oh for Christ's sake!
[Silence]

I lie still, trying to slow my breathing. After a while I begin to play back our last exchanges in my mind. Attack him, go for him . . . just like all the other times . . . Suddenly I see the barbed wire of my untold dream.

Then I think about my name. I am Barbara, Barb . . .

Then, to my great surprise, I burst out laughing.

I heard you! I was attacking you! But I heard you! [I'm not sure
*   why this is so funny.]*
Oh? Well . . . that's good.

Afterwards I write in my journal, 'I believe I will get through this crisis and through this illness, and come out changed.'

In the months that followed I did some things that seemed to show that I was getting well. These things generally went badly, as if to demonstrate the hopelessness of my case.

I received the money from my share of the house. A friend went flat-hunting with me and we found a nice little place a few blocks away from the hostel. I didn't have quite enough money for it. Mum and Dad gave me the remainder and I made an offer, which was accepted. This precipitated a series of panic attacks so overwhelming that I almost withdrew my offer, and it was only friends' interventions that kept the purchase afloat.

My friend Max held a small fortieth birthday party for me. I had a panic attack shortly before the guests arrived and ended up sitting glumly in the kitchen while my friends tried to make conversation around me. 'More like a funeral,' Max grunted afterwards.

I went to visit friends in Toronto during the Easter break. This was a really foolhardy expedition. I was panicky most of the time, incapable of enjoying my friends' company. At the end of a terrible afternoon I spoke on the phone to a Toronto psychiatrist, who clearly thought I should be in hospital. He prescribed me more drugs and signed off on a warning note, 'You sound in big trouble.' 'No, no,' I said, 'I'll be OK once I get back home.' (Don't try to lock me up here!) I told my friends that things were getting better. They were too kind to show their scepticism, but later one told me that she had thought she might never see me again. 'You seemed half dead so much of the time.'

In fact I was coming to life. While I was in Toronto I had a dream that showed me this (although I didn't understand it at the time), and which still makes me smile when I recall it.

I dreamed that I was looking for an essay by the psychoanalyst Donald Winnicott.[1] This essay, titled 'The Use of an Object', describes the process by which a baby ('the subject' Winnicott calls it here) learns that its mother ('the object') is a separate being, a person apart from itself. Mother–baby relationships were Winnicott's speciality and he wrote about them in highly imaginative, occasionally startling, ways. This essay is typical, showing how the human capacity for relating begins in an unconscious dialectic between destruction and survival.

A newborn baby has no concept of self and other. To a newborn's mind, Baby is everything, and Mother is merely part of Baby, under Baby's omnipotent control. But as an infant matures, this begins to change. Babies, as Winnicott repeatedly reminded his readers, are slave-drivers, petty tyrants who greet every frustration with enraged aggression. Every delayed feed, every wet nappy, prompts fantasies of total destruction: the frustrating Mother is obliterated. In 'good-enough' mother–baby relationships, Mother survives these imaginary onslaughts (and the biting and scratching that may accompany them) by just riding them out, and going on looking after Baby without retaliating for the aggression. Each episode of destruction/ survival is greeted by the subject/Baby with relief and joy:

> The subject says to the object: 'I destroyed you', and the object is there to receive the communication. From now on the subject says: 'Hullo object!' 'I destroyed you.' 'I love you.' 'You have value for me because of your survival of my destruction of you.'

As this process repeats, Mother gradually becomes part of external reality. Infantile omnipotence gives way to Baby's recognition of its reliance on a real and separate person 'out there in the world'. Affection is born, and concern for Mother, whom Baby can now 'use' – that is, relate to – as an independent being, and through this move forward into healthy maturity.

In most mother–baby relationships, this 'vital phase' of psychological growth is completed without too much pain. But Winnicott sounds a cautionary note. This phase is the 'most difficult thing in human development', he says – so difficult that it sometimes fails. When it does, the Baby may end up going through it much later, in a psychoanalysis. Now it is the analyst who must survive the patient/Baby's infantile destructiveness and thus become a separate person, otherwise the patient may remain stuck in what is effectively a self-analysis. 'The patient may even enjoy the analytic experience but will not fundamentally change.' But this stage of an analysis can be very risky, Winnicott warns. If the patient is too destructive, if the analyst is not tough enough to withstand the patient's aggression, the whole process can implode. 'These are risks that simply must be taken by the patient.'[2]

I had read this essay some years earlier. I knew I had found it interesting, but I couldn't remember what it said. In my dream I looked for the book containing the essay, but I couldn't find it anywhere. Awake and back in London, I still couldn't find the book because it was stored with the rest of my possessions. 'So I don't know what Winnicott says there,' I told V. He chuckled knowingly, which irritated me. 'I'm not surprised you don't remember – too close to home . . .'

As if to prove his point, my attacks on him ramped up. In the weeks that followed I smashed away ruthlessly – at him, at me. My insomnia, always acute, worsened until I was lurching about on two or three hours' sleep. I spent my nights cataloguing everything I had been through, and promising myself worse to come. The Laughing Woman screamed her delight. I decided I should go back to Friern. 'Lock the bad girl up again?' V said. Well, why not?

At the end of May my flat purchase went through. A friend collected the key for me; I couldn't set foot in the place. The following week I had panic attacks during my sessions on Monday, Tuesday and Wednesday. Thursday I arrived announcing that I expected another attack imminently.

Oh? Well . . . If you are determined to destroy yourself I can't stop
  you.
What? What??
You heard me. I'm doing what I can. But I can't stop you from
  attacking yourself, or me, or the analysis. You are trying to

destroy the work we are doing here. And you can do that, if you are determined enough.

I'm doing everything I can! I can't stop the panic attacks! I don't control them!

Not consciously, no. But you are so destructive . . .

I can't help it! I don't know what to do! What should I do? What should I do?

I am just telling you that your destructiveness has effects, has an impact.

On you?

On me, on you, on the work we do here. I'm just trying to point that out to you.

What can I do? What can I do?

I am not asking you to do anything. I am just trying to show you what is happening here.

When did I finally begin to hear him? Was it that morning? Had it been the morning a few weeks earlier, when I had laughed with pleasure at an interpretation? Had part of me always heard him?

The next day I arrive feeling calm and friendly. I tell him a dream; we chat a bit. Then he says:

You find it hard to accept that I pay more attention to you than your mother did.

What?

I said, you find it hard to accept that I pay more attention to you than your mother did.

As he repeats this I feel a movement inside me. Something inside me shifts to one side – my heart? Later I think it had felt like I was sitting on a bench, and had shuffled over to make room for him.

I can feel you here. I mean, I know you are here. I know you are here with me.

Oh? Good.

A few days later my right foot began to feel strange.

My right foot feels strange.

Strange how?

Like the muscles are moving all by themselves. Like there's something going on in my foot, I don't know . . . maybe feelings?

Feelings?

I woke up with all these feelings . . . sensations . . . they went down, into the muscles . . . It feels very strange.

Your right foot?

Yes. [I wiggle it to show him.]

[He laughs.] *As far away from me as possible.* [His chair is positioned behind my left shoulder.]

Hmm . . . yes, I suppose so. But why?

To protect the feelings from me? That might seem sensible.

No, no, I don't think so.

Well, then, a mystery. We'll have to wait and see.

24

Cure

The sensation in my right foot signalled a metamorphosis in my symptoms. The panic attacks stopped abruptly and my drug intake plummeted. My waking state changed from numbness or pain to violent agitation. Every morning I surfaced on waves of excitation. The hours before my sessions were spent wrapped around a pillow to stop myself blowing apart. After my sessions I would sometimes lapse into strange dreamlike states, neither asleep nor fully awake, cradled inside a soft cocoon of sounds and images, aware of my surroundings but disconnected from them. These reveries could last for hours. People at Pine Street got used to seeing me stretched out on the sofa. One day I woke to discover that someone had put a blanket over me. 'That looks comfy,' Nick said as he passed by.

My daily rhythms changed. Leaving Pine Street at closing time, I no longer went to Anna's house but instead returned to the hostel and placed myself in front of the TV. Every weekday I did this, sitting for hours, watching whatever passed before my eyes. My housemates came and went. My new behaviour puzzled them. 'OK, Barbara?' I bought takeaway dinners for myself. I love Chinese roast duck. A Chinese takeaway around the corner from the hostel did very good roast duck. Once a fortnight I went and bought some. Then one week I bought roast duck on Monday and again on the following Saturday. Too much! When I told V he laughed. 'So much duck! Who knows, maybe you'll even start enjoying life.' I laughed too, and again felt the sensation of moving over to admit him. 'I know you are here,' I told him.

I go to see Dr D, who is very pleased with me. No booze, fewer pills. 'It sounds like things are really changing, Barbara.' I agree politely but I am wary. I don't trust myself. I don't hear the Laughing Woman, but I fear she may be lurking about.

It's late July, three days before the summer break. I come to my session with two dreams, both very brief, just a few images. The first is of a small dog wading in a river. In the second I am trying on a blue dress that doesn't fit me. I am in a foul temper, very sullen. Fantasies about leaving analysis are darting about; I want to stab V with them. I grumble out a few associations to the first dream and then wait impatiently for his interpretation. But he isn't cooperating; instead of an interpretation I get a question.

What about the other dream? The second dream?
What?
I said, what about the other dream? No associations to that dream?
I don't know, maybe, I haven't thought about it. What about the
*   first dream? What about what I said about that?*
There's nothing there.
What? What do you mean?
I just told you. There's nothing there. You've told me nothing.
Nothing! What do you mean, nothing?!
Just what I said – there is nothing!
What? What?
I don't know why you don't understand! Listen to you . . . you're
*   not even trying! I don't know why you can't do this!*
What do you mean? Do what? I'm doing my best here!
Yes . . . yes . . . [Long pause] Why are you here?
What? What? What do you mean?
[Suddenly V seems to have had it with me.] *Dreams! I have tried*
*   and tried to show you how this works, how the mind works. But*

you don't want to learn anything; you just want to go on hating me, attacking our work here, attacking yourself. I don't know what to do; I don't know how to help you.

He draws breath and stops abruptly. I go cold.

I don't remember the rest of this session. Afterwards I staggered to Pine Street and collapsed on to the old sofa. My heart was beating so fast that I felt dizzy. A fish tank stood on a stand next to the sofa. I lay for a long time watching the little fishes circling near the water surface. Slowly my heartbeat steadied and I felt myself sinking into reverie.

A sudden jolt. I sat up.

A movement, a shift . . . My heart began to thud.

A tension, a movement . . . I felt myself go rigid, I stopped breathing, my heart was thundering.

A movement, a release . . .

My mind opened, and my neglected dream rushed in. A flood of memories, images, sounds: blue dress, sky-blue, the sky outside V's window, an open window, shouting below the window, a woman is shouting . . . On and on they came, wave after wave . . .

Excitement cascaded through me. My dream! The dream that I had lugged to V tight-wrapped like a Dead Sea scroll, for him to read the hieroglyphs, as I had brought all my dreams to him over the years . . . It was mine, my living creation! My mind was a flow, a mnemonic tide, awash with vitality. I throbbed all over with the thrill of it.

I jumped up and grabbed a piece of scrap paper from the table and wrote everything down. I was panting with emotion. When I finished I collapsed back on to the sofa. Sensations were hurtling around inside me; I couldn't stay still. I got up and moved over to a chair on the other side of the fish tank. 'You all right, Barbara?' a voice said, but I couldn't reply.

I sat and watched the fish. Slowly my mind quieted. A goldfish peered out at me.

Then an image took shape in front of me, so corny that it made me grin. My face, V's face, separate but overlapping, floating above the fish tank like a Valentine hologram. There he was, there I was, there we were – 'working together'. Look at us!

Joy flooded me. I was so delighted, so happy, I had to share this with someone. I rushed over to Gladys and told her all about it. 'Sounds nice,' she said bemusedly.

The next morning I lie on the couch reading to V from my piece of paper, tingling with my discoveries, with what has happened to me.

> *Good. That's good.*
> [Is that all he is going to say?] *'Good'! – is that all you're going
>   to say?*
> *Yes, very good . . . And now you roll over and burp and go to sleep.*
> *What??*
> *I said, now you burp and go to sleep.*
> *How silly!* [How silly, how disappointing, how annoying! The
>   old rage swells . . . and then subsides. I laugh.] *So, just like
>   that . . . a well-fed baby.*
> *Yes, that's right.*

My empty world was no longer empty.

It would be nice to say – and that was that. And in some respects it was. I had turned the corner. But V and I had twelve more years together, and they were often tough. The summer break that began the day after this session started off very badly, with a week of wretched self-loathing. The Laughing Woman still had some jokes to play. I told myself that all my unhappiness was mere self-indulgence: how could I make such a fuss when there were starving children in the world? I was preoccupied with this question.

'Starving like you were starving?' Cora – no slouch in such matters – inquired when I told her my thoughts. 'No, really starving!' I snapped at her.

But in the course of the break I began to feel much better, and when V and I met again in September it was on a new footing.

'So now you can start doing psychoanalysis,' Cora observed. V liked this – 'clever friends you have!' – but I was irritated.

So what the hell does she think I've been doing?
Living in hell? Hard place to do a psychoanalysis.
Hmm . . . yes.

At the end of September Anna went off to her teaching in the United States and I took up my room in nurse Shirley's house. I liked Shirley. She fed me huge roast dinners and gossiped about her ex-husbands. Her house was warm and clean. I went there regularly on weekends for a few months, but after Christmas my visits tapered off and by the following Easter I had given up the room. I didn't need a caretaker any more. This was a liberation so profound that it was months before I really took it in.

And the deadly oneness that had been V-in-me was gone. In its place was a living connection, very new and strange. Tentatively I tested it out, waiting anxiously for his reactions before trying anything further. A separate mind, a not-mine mind, was at work behind his words. Instead of terrifying me, this recognition gradually emboldened me. 'Ho! Revolution!' he exclaimed one day when I queried an interpretation.

Well, am I right?
Hmm . . . and if I say you are? Where's it going to stop? You might
  start thinking for yourself! Is that allowed?
[For some reason his pleasure irritates me.] *You sound so pleased*
  with yourself!
And what's wrong with that?! I've worked very hard to get you here.
[This is true, and I soften.] *Yes . . . But you know, I've worked*
  hard too.
Yes! But you were so frightened before.

I wanted to be good, I thought if I was good I would get better.
Hmm . . . you weren't being 'good', you were being compliant. Out
    of fear. Except of course when you weren't. Now you're learning
    to cooperate. That needs two people.

Slightly to my surprise, I understand this completely.

It was not only with V that I experienced such changes. Betsey was one of my Pine Street chums. One evening she and I went to my Chinese takeaway. By now the owners recognized me – 'Roast duck, yes?' We sat down to wait. On a bench across from us a young man was feeding bits of food to his girlfriend, who bridled and cooed her appreciation. Both were dressed in motorcycle gear, all heavy leather and metal studs. Betsey caught my eye and we grinned at each other. I was astonished by this.

Why did it surprise me so much, us smiling together like that?
Because you felt the connection between you? You're very fond of
    Betsey.
Yes! I felt it, the feeling between us. It was like a little spark.
Yes. Lovely feeling, isn't it?
Yes . . . yes . . .

The next night I dream that I am in a basement room like the one in my childhood home, but infinitely larger. The room is full of fridges and freezers – they line every wall, they stack up to the ceiling, there is hardly room enough to squeeze between them. I go from one fridge and freezer on to the next, peering inside them. Every single fridge and freezer is crammed full of roast duck: freshly roasted duck, leftover roast duck, frozen roast duck, an endless supply of delicious duck – all of which has been put there, I suddenly realize, just for me. I laugh with delight, and wake myself up.

25

Stories

I have uttered my life in my own voice.

J. M. Coetzee, *In the Heart of the Country*[1]

My career as a mental patient was drawing to a close. Its final phase saw me testing out my adulthood. I had spent so long as a dependent child-adult; could I do grown-upness for real this time? I began cutting back my time at Pine Street, spending whole afternoons and evenings at my flat sorting out my things, planting window boxes, cooking meals for friends. I returned to the hostel only to sleep. This was good; I was remaking my life. But it wasn't enough; I needed to work. Not research or book-writing . . . that still seemed a long way off. Maybe some teaching? I remembered a woman coming to the day hospital to tutor patients with reading difficulties. Maybe I could do this? I knew there were people at Pine Street who could barely read, and I asked Nick if I could offer help. 'Why not?' I collected some teaching materials from a local college and announced my availability. People were surprised – 'You're a teacher, Barbara? Why are you here then?' – but two signed up.

Gail was a wild woman – frantic one moment, radiantly joyous the next. We had one lesson, which ended when she chucked a notebook at me. Nigel, on the other hand, was quiet, self-contained, diligent. 'I hope my voices let me do this,' he said when we began. 'I hope they don't start shouting.' We met twice-weekly for an hour and made slow but discernible progress. Encouraged by this, I went along to other psychiatric day centres in north London to offer my services. Nobody wanted literacy tutoring. How about creative writing? This met with more enthusiasm, and by May 1991 I was teaching classes in two day centres (both, like Pine Street, now closed). I began to see earning possibilities, and signed up for a three-weekend training course in teaching adult literacy.

I still have a file of papers from this phase of my teaching life. Looking back over them, I find that for Assignment Four of the literacy course ('Discuss some of the challenges of teaching literacy to adults') I chose to describe my work with Nigel (using a pseudonym for him of course, as is 'Nigel'). 'Ken tends to read word-by-word with little sense of the overall meaning of what he is reading,' I wrote.

> Sometimes he does not realize that his reading of a word makes non-sense of the phrase or sentence in which it occurs. I think this is largely a matter of attitude – Ken is so used to no one paying attention to what he says, it seems to him that he could spout any nonsense and it would make no difference . . . Ken is not used to people treating with respect his attempts to express himself meaningfully.

Ken/Nigel was not alone in this. I had now spent three years in the company of people with severe mental illness. Had I always 'treated with respect' their efforts 'to express themselves meaningfully'? Becoming a teacher placed me in a new relationship to people's words. It also gave me a degree of power over my fellow patients which confused me for a while, in this mid-life transition towards adult responsibility.

My qualifications for teaching creative writing were slim. In the early 1980s I had published two children's books and a poem – not much to draw upon. But I was sanguine. I'd always been a storyteller; storytelling was easy. How hard could this be?

Madness has long been linked to literary creativity. The 'fine frenzy' of writerly inspiration is one of those clichés that litter popular representations of mental illness. Like all such glamorizations, its flipside is contempt. For every John Clare or Sylvia Plath, there are tens of thousands of ordinary 'nutters' who are regarded with disdain. And like all clichés, it obscures a more complicated reality: on the one hand, the hard discipline and craft of writing; on the other, the richly imagined narratives of ordinary madness. I saw plenty of evidence of the latter during my time in Friern. Ada with her sky full of star-babies, Magda's Iranian water-poisoners, Fiona's spaceman with his sojourns on Saturn . . . stories like these had been everywhere on Ward 16. One day a young woman, new to the ward, had kneeled down in front of me to describe her love affair with an Oxford don. She sat at my feet, clutching my hands, telling me about the flowering of their romance, the wicked conspiracies to keep them apart, the secret messages her lover imparted to her in his lectures, the evil people who had forced him to deny his passion for her, and even, on one terrible occasion, to call the police when she came round to his rooms . . . She was distraught but wonderfully eloquent – a gothic novelist manquée. I had never encountered erotomania before and was fascinated, as I was by Nora's stories of corporate malfeasance and the galactic adventurers of Fiona's boyfriend.

These were marvellous inventions. But vivid and elaborate as they were, tales likes these were not crafted creations. Such fantasies entered people's minds unbidden, like wide-awake dreams or night-mares, holding their narrators spellbound – just as my fantasies about my past had held me spellbound until V and I dismantled them. The men and women who recounted such stories to me were in thrall to them, often tortured by them. Yet now I intended to ask my students to deliberately fantasize, to 'make things up' as if this were just a bit of casual fun. For some people this was, quite literally, a crazy sug-gestion – but understanding this took me longer than it should have.

My two classes were small, about a half-dozen students each, mostly 'revolving-door' mental patients – that is, people who had been in and out of hospital many times. Everybody could read and write, although some with difficulty. I didn't ask for diagnoses and the day centres didn't brief me. But most of the students were forthcoming about their ailments, talking freely, occasionally obsessively, about medications and swapping tales about hospital experiences. Here was a rich seam to explore! It didn't occur to me to suggest this. Instead I devised a set of exercises to get the creative juices flowing. Week after week I brought in teaching aids – magazine ads, an array of small objects, a few celebrity photos – and asked the students to write stories inspired by them. They set to with a will, but the results were disheartening. Most managed only a few laboured sentences. We tried out some group exercises and this worked slightly better, but I was discouraged. I cast about in my mind for new stratagems.

How about 'guided visualization'? I had done this several times at Pine Street, listening with closed eyes as one of the centre's many therapeutic apprentices took a group of us around some imaginary scene, leaving us eventually to complete the fantasy ourselves and then, if we chose, to write it down. I had found the exercise potent – too potent on one occasion, which I now conveniently forgot – and thought it worth trying. So: 'Imagine yourself in a big room, with a bed and a door. You've just woken up. You get out of bed and walk over to the door. You open it . . . What do you see out there?' Everyone had their eyes closed and things seemed to be going well until Doreen, a middle-aged woman with a strong Geordie accent, suddenly exclaimed, 'My glasses! My glasses! What about my glasses!' She couldn't go outdoors without her glasses. 'I won't be able to see! Can't see without my glasses! There's no shelf for my glasses . . .' I tried adding a shelf next to the bed ('See? Your glasses are right there! You can put them on as soon as you get up.') but she was much too upset to absorb this and her distress affected everyone. So much for guided visualization.

So I tried dreams. Most people professed not to remember their dreams but a few said they did and agreed to have a stab at recounting them. Most of the resulting descriptions had an off-the-shelf quality ('I dreamed I won the pools'; 'I dreamed I was on a sandy beach') but not all. A woman in her sixties described a dream in which she was attacked in her home by a man in a flat cap ('my father always wore a flat cap'). Andy, a gentle, anxious man who lived with his mother and a canary, recounted a 'terrifying' dream involving his mother; he wouldn't say what had frightened him so much. 'Come on, Andy, tell us what happened!' a man named Ron said, but Andy shook his head. The following week Ron wrote about a dream in which his entire family died in a house fire. He came up to me afterwards. 'If they die like that, it will be all my fault!' 'But it was only a dream . . .' I said feebly. He shook his head dolefully. 'No, no, it will be my fault!'

I kept most of the students' compositions, and reading these dream narratives now I am shocked by my recklessness. But I had an agenda. Although I wouldn't admit it to myself, and certainly not to V, I was playing at psychotherapist. Therapists – V, Dr D, the Pine Street therapists – were my grown-ups now; I wanted to go over to their side. I contacted a psychotherapist whom I knew slightly and asked her to supervise my teaching. (Trainee psychotherapists, when they start seeing patients, are supervised by senior therapists.) She agreed and we met regularly for several months. I was frank with her about my situation but I didn't admit to either of us the fantasy that I was enacting. But V saw what was going on and eventually we had a thorny exchange about it.

> *I've been wondering about maybe becoming a therapist . . . just*
> *thinking about it, you know . . .*
> *I take it this is just a fantasy!*
> *Why? Why shouldn't I think about becoming a therapist?*
> [Silence]
> *Yes, well, OK, maybe not . . . You know, when I was a child I*
> *thought I might become a vet . . .*

Hmm . . . well, some people say animals have an unconscious. Stick to teaching.

So I taught, and learned. Over the months people became more confident and I began to see changes. Andy, who at the start had expressed strong anxieties about his writing abilities, began to carry a notebook for his jottings. A young woman named June wrote poems that she turned into songs. She and her friend David sang some of these songs in class. Colin, the breakfast-club organizer from Ward 16, turned up midway through the year and wrote sheaves of poems. 'Very Wordsworthian,' I said about one, which turned out to be just what Colin wanted to hear. 'I love Wordsworth!' he said, and proceeded to recite the opening lines of 'I Wandered Lonely as a Cloud'. These were pleasing developments.

But it was Ron who really surprised me. I had been with Ron in Friern and again at Pine Street. He was an unprepossessing man in his thirties, an alcoholic with long greasy hair and rotten teeth. At Pine Street he had alternated between ingratiation and aggression and been widely disliked. 'Slimy', had been Betsey's summary description of him. 'He makes my skin crawl.' My heart sank when I first saw him in my class. But he wrote copiously; I have a small stack of pieces by him. At some point (presumably before V put a check to my therapeutic aspirations) I asked people to write about their memories. Ron wrote about visiting his parents in Devon. He described waking up one morning, desperate for a drink, fishing a can of beer out of his suitcase. 'Before I opened the can I drew the curtains. I saw my mum out in the garden picking some vegetables for lunch, my father came out to speak to her. I felt so terrible been [sic] on a binge with the Alcohol, drinking my beer and seeing my parents from the window. It made me feel unnatural.' He wrote about this scene several times. He also wrote about living rough and begging. 'I know it isn't right to beg but I got very hungry sometimes. I spent the money on beer tho.'

A few weeks before the class ended I wrote some notes about it for myself ('in case I ever want to write about this'). I described the changes I had witnessed in Ron, as the unsavoury man I knew at Pine Street gradually became a lively classroom presence, 'chatting, responding to people, being listened to by them. I think of this as a real accomplishment – both for him and for me'. I suspect I flattered myself here. In fact Ron was on the wagon over this period: hence his increased vitality and responsiveness. Nonetheless this was an important reminder of the satisfactions of teaching, which after the first year had begun to wear thin for me.

At the beginning of 1992 I persuaded the day centres to pay me for my classes. I was pleased to get the money, but being on the payroll somehow made the teaching feel more routine, even boring. And my identification with the students was fraying. A bossy woman named Eleanor baited a pretty young girl constantly, making the girl very miserable. I didn't know how to stop this and it frustrated me. Dan, a bolshy man in his twenties, arrived in my class one day drunk and rowdy. I spoke to the manager, the police were called and Dan was carted off. Before he left he insisted on returning to the class to assure me that he would be back soon. When I returned to the centre the following week there was a letter from Dan accusing me of turning him in to the police and threatening to 'get me'. Nothing came of this but I was frightened.

Another student was a psychiatric nurse who worked part-time at the day centre. Her writings were very violent, full of blood and corpses. At first I had found this mildly amusing but now it disturbed me. I thought of speaking to her, but what would I say? Then one day Ron made a ribald remark to me and I told him to fuck off. This was not good. I had had enough, time to stop. I taught on until the summer vacation and then drew the classes to a close.

By then I had also finished at Pine Street. My lessons with Nigel had ended a few months earlier and I had no reason to go to the day centre. Nor did I need nurse Shirley's room any more; I was living full-time in my flat. I thanked Shirley and we parted. I hung on to my hostel room over the summer break, 'just in case', and then gave it up in the autumn. One October afternoon I went to see my GP, whose surgery was around the corner from the hostel, to tell him I was transferring to a GP practice nearer to my flat. He took a letter from my file. 'You've been discharged,' he said to me. 'Discharged?' 'I've received this from Dr D. She has discharged you. Congratulations.'

> *I'm discharged. Dr D has discharged me.*
> *Oh. Well. Is that good?*
> *Yes, I think so. I mean, it makes sense.*
> *You don't need her any more?*
> *No. Well . . . I don't know, I don't think so.*
> *You're not sure?*
> *Sure?? How can I be sure? How can I ever be sure about such a thing?*
> *No, well. But perhaps sure enough for now, to manage without her?*
> *Yes, sure enough for that. To do without her.*
> *But not without me. Not yet.*
> *Is that a question?*
> *No.*

26

Endings

Life post-discharge felt both familiar and strange. 'I'm like a building under reconstruction,' I reflected in my journal, in a series of entries looking back on my illness. 'Same old materials, new foundations.' I had many things to do, and to redo. I had a new sense of urgency, and time began to blur again. For nearly a decade, my days had proceeded with excruciating slowness. I had watched other people hurrying along, worrying about the time, hoping that their watches were fast, while I – as I recalled in my journal – had experienced my minutes and hours as an interminable succession of

> relentless *details* – the endless remorseless detail of daily
> survival. I remember feeling that so often when I was crazy –
> how I had to put on those clothes, walk down those stairs, get
> through these thousands of little movements and changes of
> the day – each of them so intolerable, taking so long . . .
> minutes that were hours, hours that never ended . . . Now
> things blur – into a preoccupation with ideas, goals, events,
> desires. When I had none of those, there was nothing but the
> small drip-drip-drip – Perhaps this was one reason why I drank.

So now I was getting busy again. 'Is that a briefcase?' V said one day. 'Yes, I'm going to the British Museum.' I had just received a publisher's letter inviting me to write the Introduction to a bicentennial edition of Mary Wollstonecraft's 1792 magnum opus, *A Vindication of the Rights of Woman*. I had tossed the letter away when it arrived, as I had thrown away all such invitations over the years. But then I had retrieved it from the bin, and now I was off to work.

The Introduction was published in autumn 1992. Eighteen months later I travelled to North America to give some talks on Wollstonecraft, and on my return I was appointed to a lectureship in history in the Department of Cultural Studies at the University of East London.

Monday 17 May 1993. I go to my session knowing I have a job. I tell V and he whistles with pleasure. We chat about the teaching, my research. All is friendly and happy. Then V says, 'So, where's the poor mad woman now?'

> *What? Oh . . . Oh, I don't know . . . I guess she's still around somewhere.*
>
> *Yes, I think so. She was you, after all. And you mustn't forget her. She saved your life.*
>
> *Saved my life? Really? How? What do you mean?*
>
> *What do I mean . . . ? Well, when you came to me you had three options. You could decline into a psychosomatic illness from which you'd probably not recover. You could become a long-term psychiatric patient, ending up in so-called 'community care'. Or you could descend so deeply into alcoholism that you'd never be retrievable. You tried out all three. But in each case there was something in the way that you did them that saved you . . . that kept you alive. So your craziness was destroying you but also keeping you going, both at once . . . Do you understand me?*
>
> *I don't know . . . I think so . . . You mean, like the way I thought about the booze? I always thought I couldn't do without it . . . The pain was so bad, without the booze I couldn't keep going, I would have killed myself. But the booze was killing me . . . killing me and keeping me alive . . .*

Yes . . . it was always a balance . . . The crazy woman was balancing there, between life and death . . . and she is there still, although nothing like so vicious now. We must not forget her, or repudiate her, because of what she did for you. [Long pause] *And also because if you try to forget her she will retaliate somehow.*

What? What do you mean? You mean I'll go crazy again??

No. But you'll pay a price. [Long pause] *You've worked very hard . . . and sometimes the crazy woman has worked right along with you.*

Yes . . . [Long pause]. *You know, I was thinking about Mum the other day. I thought something I'd never thought before. I thought, that when I was young, when I was a girl, before I came to London, I thought that part of Mum, part of her was yelling at me – 'Run away! Run away! Get out of her! Get away from me! Get away!'*

Trying to save you?

Yes . . . to save me . . .

Hmm . . . yes, perhaps . . .

I won't forget the crazy woman. How could I forget her? She's me. How could I ever forget her?

I recorded this dialogue in my journal and then I wrote: 'If this were just a story and not a messy life, it would end here. But on it goes . . .'

And on it went. I started my new job and liked it, and my confidence grew. I gave more talks on Wollstonecraft, acquired a cat (whom I named Blue, after the dog at Pine Street), worked on my book. But I was still waiting for my miracle. My anxiety levels had plummeted but my nights were turbulent and I woke often in pain. I was still plagued by self-dislike, and I was lonely – not with the terrible nullifying solitude of my crazy years but with a continuous ache of loneliness that made my days feel parched and grey. I felt old, ugly, contemptible. No one would ever love me; I would spend the rest of my life alone in my flat with a cat for company. This was to be my punishment for a host of crimes: envy of others, especially V; forbidden sexual wishes; aggressive fantasies, some very violent . . . all the standard fodder of a common-or-garden psychoanalysis which I was now, finally, doing, although not without the occasional backward step and plenty of grumbling.

I took a summer trip to Saskatoon. I felt sure that I was well enough to go – and I was – but the return was much stranger than I anticipated. I had been thinking about it, dreaming about it for so long – this little prairie metropolis, my 'home town'. In my dreams I had seen it not as the pleasant provincial city Saskatoon was, and is, but as the small child had experienced it: as a place of heart-scouring loneliness, of fear and desolation. And I was not meant to be back here. Suddenly I knew this. I was not supposed to have made it back here. I was supposed to have done what I had so nearly done: gone away to the big city, covered myself in glory, and perished. I had flouted my fate. For several days I felt like a ghost.

I didn't stay with my parents on this trip but I spent a lot of time with them. I know that I did this because my journal says so, but I remember nothing about it, except for one episode, which I recorded in my journal. Mum had a hospital appointment for a needle biopsy on a breast lump. I offered to accompany her, which surprised her. It proved to be a cyst; all was well. I was relieved and said so. Then we went shopping

> . . . and suddenly she reaches over and takes hold of my chin. 'Smile, Barb – aren't you pleased for me? Aren't you glad I'm well? Why don't you give me a big smile?' Right out of the blue – accusation, guilt, coercion. God forbid she could ever take in any concern or affection from me that she hadn't bludgeoned out of me.

My journal added that I felt 'repressed, bitchy', but that things were 'manageable'.

The following week, back in London, I dream that I am visiting Magda in hospital. She is sitting propped up against a wall. Her hair has gone white and she looks small and vulnerable, and very mad. I take her hand. 'Hi, Mum,' I say. She doesn't reply and we sit on in silence.

My lack of memories from this Saskatoon trip was not incidental. For years I had tried to blank out Dad and Mum, to banish them from my life. I had no birth parents any more; V was my parent, Dr D was my parent; they were my real mother and father. These fantasies persisted long after I began to lead my life again.

But Dad and Mum, especially Mum, were not erasable. They made their presence felt, not least through their financial support. I relied on this money but I wasn't grateful for it. I loathed taking it from them – not because I yearned for adult independence, but because I didn't trust them to go on providing it, and because it got in the way of the unadulterated hatred I wanted to feel for them, or rather the hatred I *did* feel, mixed with the terrible guilt of the greedy, thankless child that I knew myself to be. And the money had a price. I had always to ask for it and was regularly reminded that supplying it was a burden to them, as of course it was, although I was reluctant to acknowledge it – a bit of psychological foot-dragging that occasionally irritated V (who was also subsidizing me by charging fees well below his usual rate).

> *I don't want anything more to do with them! They're not my parents . . . no real parent treats a child like they treated me. Why should I have anything more to do with them?*
> *They support you. With money, I mean. What about that?*
> *Oh that!*
> *Yes, that! They have given you a lot of support that way.*
> *Yes . . . well . . .*
> *Yes! You shouldn't ignore that. You owe them a lot.*

Psychoanalysis was beyond my parents' ken. Like most

left-wingers, Mum and Dad had read some Freud. But psychoanalysis as a clinical practice was unknown to them, and my fervent commitment to it puzzled and worried them, Mum in particular. Mum was, as she always had been, the parent in charge. And I was the child of her soul, her younger self. She needed to know what was happening to me. This was understandable, especially as she was helping to pay for it, but her intrusiveness terrified me.

'So tell me more about this psychoanalyst, Barb . . . What do you talk about with him?'

'Oh! Oh . . . lots of things . . .'

'About me and Dad?'

'No, no! He's not interested in anything like that . . . no, no, no . . .'

'Oh really? I thought maybe that was what it was about . . . ? Your childhood and so on?'

'Well, a bit of that, yes, but mostly just fantasies, you know . . . dreams and things . . . just stuff like that.'

I never told Mum V's name, despite her repeated requests for it; I was afraid she would contact him. But she knew about Dr D and wrote to her several times. She also went to see her once, and seemed to find their conversation reassuring. This was on one of several visits she and Dad made to me over these years, including one during my last stint in Friern. I didn't want them to come into the hospital so we met for lunch at a café near the main gate. 'It was so sad to see her there, my beautiful, clever daughter,' Mum wrote later that day, in the diary I saw after her death.

On another visit Dad wept. I had gone to have dinner with them at their hotel, feeling and looking dreadful. The three of us had sat in the restaurant saying little. Suddenly Dad rushed from the room, returning a few minutes later, red-eyed. Later he sent me a note apologizing. 'I am so sorry, Barb, that I lost control like that. It was wonderful to be with you.'

They loved me as best they could, and it wasn't good enough, and I made them suffer greatly for it. I know this now and it grieves me.

Could I have done things differently, behaved more kindly, tried harder to move our relationship on to a better plane? The question cannot really be asked of my madness years, but later? I would like to think so, if there had been enough of later – but at the beginning of 1995 Mum had her stroke and everything changed.

The stroke was a severe one. Mum was comatose for three weeks and incapacitated for nearly a year. Over the next three years she recovered to the point where she could live semi-independently. But she was much aged and changed. The tense, assertive woman I knew had been replaced by a childlike old lady, disinhibited, sweet-mannered, quick to express hurt or affection. The transformation disoriented me. 'My mother is gone,' I told V after one of my visits, 'yet there she is.'

She was also a much happier person. She moved into a supported housing complex where she enjoyed the company. 'How are you today, Mum?' 'I feel wonderful! So much to do here! So many nice people!' Her old preoccupation with 'importance', with public reputation, persisted, but it no longer soured her everyday pleasures. And her cheeriness warmed me up too. She delighted in my job and my books (which she sat over, hour after hour, although she could hardly read) – and I was glad of this. I even found myself reminding her of her own claims to 'importance' which she had forgotten. I showed her photos of her at various official functions. 'That's me? Well, well . . .' She did not remember anything about my illness, but occasionally she looked at me with a slight puzzlement. 'You know, Barb, you seem different somehow than you were . . . ? But maybe that's just my funny memory, my illness, I mean?' 'No, Mum, you're right. I was sick and it changed me a bit.' 'Oh yes, of course . . . that must be it. Well, dear, you seem very well now.'

So, in a final serendipitous twist to our long shared history, the very different illnesses that almost claimed Mum and me took us into better lives, apart and together.

Dad's last years were much tougher. Mum's stroke knocked the ground out from under him. Mum had always looked after him, far more than I realized, and without her he crumbled. He had been showing signs of dementia, and now these developed rapidly. He began drinking heavily and became confused and paranoid. He festooned his bedroom door with 'Keep Out' signs. 'Government agents' were out to get him, he told us. (In fairness to Dad, this had once been true, back in his Communist days.) He had some hunting rifles locked in a basement cabinet, and one afternoon he barricaded himself into his room and threatened to 'get out the guns'. He was wild with fear. 'Why is no one looking after me?' he bellowed at the women employed to care for him and Mum. Everyone, even Mum, was afraid of him.

Finally he was admitted to a secure ward in an old people's residence. By this time he no longer recognized anyone except Mum. But being looked after properly brought his anxiety levels down and his mood improved for a time. I went to visit him and found him bright-eyed and loquacious. He talked about Spain and coal-mining. On my next visit he flirted with me. He was sitting in a wheelchair, looking spruce and relaxed. He patted my hand and smiled into my eyes. 'You know, my dear – I'm very sorry, I cannot recall your name – I feel quite guilty about my relationship with my wife. I don't think I have given her much pleasure, sexually you know, I don't think I'm quite right for her . . .' He tried to hold my gaze as I looked away. He was confiding, suave, charming – so utterly different from the Dad I knew that I felt oddly comforted.

Dad died in 2003. I was relieved; his final months had been dreadful. I expected Mum to grieve – she had visited him regularly over the years, and always been very tender with him. But she didn't seem to mourn him at all. If anything, she appeared to look back on her marriage with distaste. 'Oh, why did I stay with him . . . ?' she said to me suddenly one day when we'd been looking at old photos. 'I left him once, you know . . . I should have stayed away!' I'm surprised by her mention of this. She had left him briefly in the 1970s. He had persuaded her to return by pleading the need to protect their public reputations. I knew about this from Lise; Mum had never spoken to me about it before, and never did again. But if she felt some regrets about her relationship with Dad, this was far from the whole story.

In the summer before Mum died, Lise and I were called to her bed after she had contracted a severe chest infection. She lay propped up on her pillows, wheezing, drifting in and out of consciousness. Then suddenly she cried out, 'I must get George, he's waiting for me! I must go get him right now! He needs me!'

'Dad's dead, Mum. George is dead.'

'No, no, he's waiting for me to come and get him. I must get George, I must go.'

She was so weak, she could barely raise her head. But she fought with us and cursed us as we prevented her from scrambling out of bed.

'Damn you, get out of my way! I must get him! George, George!' She was frantic.

'I'll go get him, Mum,' I said. 'I'll go get George for you.'

She stared at me suspiciously. 'Now? You'll get him now?'

'Yes, right away.'

'Go get him then!' she ordered me. 'Go get my husband.' Then she fell back on to her pillows.

Those were the last words she spoke to me. She died four months later when I was back in London.

In 1996 I fell in love with a close woman friend who also fell in love with me. This was a happy surprise to us both, as neither of us had ever been sexually involved with a woman before – but no surprise to V, or so he claimed. 'There have been hints of this . . .' he said to me complacently. My lover, my civil partner as she now is, has two lovely sons, so now I found myself with a family. Seven years later I published my book on Mary Wollstonecraft. I gave a copy to V.

Do you think you'll ever read it?
Of course!

I felt my eyebrows go up, but he didn't see this.

A few months later V and I parted. It wasn't an easy separation. I didn't want to leave. We were getting on so well, I told myself, and I still had so much to learn! He was so *useful* to me . . . why should I go? I teetered angrily on the edge of the nest.

I don't want to stop.
Yes, I know.
I get frightened . . . And I have all these dreams . . .
Yes.
I get so anxious . . .
Yes . . . Are you going to spend the rest of our time together complaining . . . ?

That was exactly what I felt like doing. I carried my complaint to my friend Adam.

'I don't want to quit!'

'No?'

'No! Why should I quit?' (I know how childish I am being, but I don't care.)

'Maybe you'll enjoy keeping your thoughts to yourself?'

Now, this is a new idea and I ponder it. Silly fantasies, embarrassing dreams . . . no one would have to know. I could have secrets! My inner life would be my own. This is tantalizing . . . but am I really ready for it?

For many years I had a recurring dream that I was living in a house without doors. No privacy, no place to go without being seen and heard. I tell the other people in the house that I want a door on my room. 'No, no,' they say, 'doors are banned here, we don't believe in doors.' (I describe this as my anti-commune dream.)

After I move into my new flat, I begin having a different dream.

Now I dream that I am in my home and it is full of strangers. I have no idea who these people are or what they are doing here. They seem to be entitled to be in my flat, but are they really? Who said so? Finally I order them out. 'This is *my* home,' I say. 'You have to have my permission to be here.'

Now, as my final session approaches, I dream again about my flat. This time I am alone, and my flat has grown immeasurably. I go from room to room, opening doors which lead into other rooms with more doors, more rooms . . . Every room has huge windows that show me beautiful views – bright skylines, sparkling water. I am thrilled, intoxicated with this wonderful infinitude of space and light, and wake up very happy.

So . . . maybe I am ready?

27 July 2003. I have imagined this session so many times. The two of us reminiscing about our long years together, friendly chat and gentle laughter and an occasional catch in the throat . . .

I arrive utterly exhausted. My partner and I have spent the previous evening driving around on Sussex roads until three in the morning, trying to find our way back from her sister's country home. I've had almost no sleep.

So, now . . . Here I am, for the very last time, here to say goodbye to this man who has changed my life. I sit in the waiting room fidgeting. Finally V calls me into his room. As I enter I look at him, and I see that on this momentous day, this final day after twenty-one and a half years, V looks and sounds precisely as he has always looked and sounded: cool, professional . . . It could be any bloody session on any damned day! I could be anyone! It is unbearable!

I flounce on to the couch and sulk furiously. But after a bit too much of this, I give up and let the sadness come through. I cry for a while, and we chat about this and that: the rose bush I gave to him earlier in the week, my summer plans. I feel the clock ticking towards closure.

You know what I am doing this evening?
What are you doing?
I'm dining with a priest and a psychoanalyst. [This is true.]
Well! So you've got your bases covered.
Yes . . .
You're telling me you can manage perfectly well without me.
Hmm . . . 'perfectly well' might not be the phrase you want.
Oh! . . . ever the pedant! 'Well enough' then. Will that do?
Yes. Well enough.
Good. It's time.

EPILOGUE:
AFTER THE ASYLUMS

Friern closed on 1 April 1993. It was a rush at the end, with a few patients still there on 31 March.[1] Remaining staff immediately began dismantling the hospital, selling off furniture and beds, emptying files on to the shelves of the London Metropolitan Archives. Four hundred beds and mattresses were shipped to Sudanese hospitals. Later, after the developers moved in, Victorian fireplaces and oak floorboards were sold as architectural salvage. In 2004, when I was refurbishing a house, I bought several metres of oak floorboards from a salvage company. I asked the salesman if he knew their provenance. 'They're from an old mental hospital, up in Friern Barnet . . . we've sold loads of them.' They are excellent boards; sanded and refinished, they bring a touch of souvenir elegance to my home.

The future of the site was still unresolved at closure. There was no shortage of ideas, with private companies eyeing up its commercial potential while community groups, heritage bodies and other interested parties put forward suggestions for a library, an old people's home, a university, low-cost accommodation.[2] A group of service users proposed making part of the old building into a museum of psychiatry. Local residents objected to some of these suggestions and ratepayers' groups weighed in. Things dragged on. A travellers' community moved on to the site and proved hard to budge. It was followed by the Metropolitan Police, who used the grounds for firearms training. Eventually part of the site was sold off for a retail park and townhouse development, and at the end of 1995 the main building and thirty acres of grounds were purchased by Comer Homes, a property development company specializing in 'historic' conversions.

Throughout these years former patients kept turning up at the site. The security men at the front gate were given information about local hospital services and instructed to turn the returnees away 'gently but firmly'. Yet still people kept coming. The company responsible for guarding the site was warned that its surveillance was insufficient: 'The patrols are frequently completed in such a short time that it would appear impossible for a proper inspection to have been carried out.'[3] I can testify to this. In April 1996, as Comer began work on the building, I came back with a friend to take a look. It was the Easter weekend and the site was deserted. We prowled around, looking through windows, taking photos. I kept an eye out for security guards but saw no one. After a while I began testing doors. My friend eyed me nervously and backed off. 'I'll think I'll just wait over here . . .' One door looked a bit wonky and I gave it an exploratory push; it moved slightly. I shoved hard and it gave way, almost tumbling me into the corridor. Hurrah! I have often surprised myself at moments like this. I am usually so timid, so nervous – but then along comes that adrenalin uprush, that 'oh well – what the hell!' and I take off. Now I was gleeful, thrilled with my adventure. I followed the corridor, skirting broken floor tiles, taking note of familiar grafitti, until I reached the stairs to Ward 16 . . . Up the dark smelly staircase and – *voilà!* here I was again, post-apocalypse: windows smashed, lockers scattered, pigeon shit everywhere. I picked my way about, opening doors, peering into empty cupboards. All was eerily quiet . . . and then it wasn't. Voices? Footsteps? I sped off – down the stairs, along the endless corridor. Definitely footsteps now. Where the hell was that damned door? Finally I found it and escaped across the ravaged lawn to my friend, giggling with fear and excitement.

Comer Homes boasts about its 'respect for history' in its Victorian conversions. But conservationists have been scathing about the transformation of Friern into Princess Park Manor: 'a Disneyland-like distortion of asylum [into] country manor'.[4] The only parts of the old hospital still intact are the facade and the reception lodge, both of which were already under conservation order when the company moved in, and fifty metres of corridor. The remaining 500 metres of corridor have vanished into the apartments. The fine old library is gone; the hospital chapel is now a gym; the graveyard (for 'pauper lunatics') has been obliterated, bar one small burial stone. Friern has been erased. All that remains to show that this was once England's premier lunatic asylum is the 1849 plaque commemorating the laying of the foundation stone, which now graces the glitzy reception area outside the gym. I saw the plaque on my 2002 visit, looking so incongruous in its new setting that it reminded me of the fake Victoriana stuck up in pubs to give them a 'heritage' atmosphere.

Today the Manor is a celebrity address, home to a bevy of pop singers, TV entertainers and football stars – the 'bling brigade' as one newspaper snobbily puts it – as well as the usual City professionals and foreign moneybags.[5] Do these residents know the Manor's history? If they do, it is not from the Comer sales literature in which the building is described as a 'Victorian masterpiece which has delighted and inspired aficionados of fine architecture for generations', along with other carefully turned phrases – 'distinguished history', 'aura of grandeur', 'a majestic profile worthy of its original classic design' – hinting at regal associations. For devotees of irony however there is plenty to enjoy, as the ads enthuse about the building's 'robust, thick, well-insulated walls' which 'inspire feelings of security and cosiness'. The Manor is 'an oasis of tranquillity from the bustle and frustrations of life', potential buyers are told. For the resident celebs, this cosy tranquillity comes with the added attraction of distance from central London. 'People like it here because they can be anonymous,' a Comer representative told a newspaper reporter.

In 1998 a film-maker came to Princess Park Manor to interview the first batch of apartment owners. She brought with her a few former patients, including one whom I knew slightly – a bright, pugnacious Mind activist. The film that resulted is riveting. As the past and present residents speak to camera, a slow movement towards each other occurs. The patients are thoughtful, humorous, fluent. One, a former theatre-worker, describes the hospital as having been 'hell on earth'. Finding himself in the gutted main hall, he abruptly opens his mouth and sings his heart out. 'I can go now,' he tells the camera. 'I'm not frightened any more.' Another, an elderly man, looks wistfully at the carcass of his ward, my 'second home', and reminisces about his time in the Friern football team. Meanwhile, a feng shui consultant is busy tapping for energy sources in the walls. 'Phew!' she cries, knocking hard. 'Something sure was going on here!' Are they afraid of ghosts? the film-makers ask the new residents. None will admit to this, but all emphasize their sympathy for the mentally ill. 'My friends say I should have been in an institution like this long ago,' one man chortles. Quizzed about their reasons for choosing the Manor, some become remarkably self-revealing, one retired man describing the 'mad' world beyond its gates as too stressful for him, while a divorcée admits that she hopes such a self-contained housing development – with its fancy leisure facilities, café and private bus service to the train station and shops – will bring her new friends and romances. She had been feeling pretty suicidal before she moved in, she confides. How ironic, my Mind acquaintance comments to the camera, that people are now willing to pay large sums of money to live in a place that advertises itself as somewhere that you 'never need to leave'. With its manned security gate, high-tech locking systems and omnipresent surveillance cameras, Princess Park Manor aims to keep out what Friern was meant to keep in: but not all the devils that beset individuals are so easily contained. Yet it is good to be reminded, as this lovely film does, of the miseries and frailties common to all humankind, whether hopefully mad or hopelessly sane.[6]

Friern's demise marked the final phase of the mental health

revolution announced by Enoch Powell in 1961. Today, all across the western world, the old asylums are history. Their former residents are back with their families, or living in group homes or social housing, or they have vanished into the netherworld of the urban homeless. People – like me – who a quarter-century ago would have been treated in an asylum are now being looked after (or not) in the 'community' through a network of services provided by the voluntary sector, private companies and, here in the UK, local NHS mental health 'trusts' (public sector non-profit corporations) which operate like private businesses, buying and selling services in the psychiatric marketplace – a situation that even Enoch Powell, with his passion for free enterprise, could surely not have envisaged back in the 1960s.

What is life like on the far side of this revolution? In the Prologue to this book I said that I wrote it with a question in mind: how would I fare if what happened to me back in the 1980s happened today? What would become of a young woman with a serious mental disorder who, having struggled and failed to keep herself afloat, found herself inside the mental health system? The rest of this Epilogue relates what I discovered by way of an answer.

Like the asylums in their heroic foundational phase, community care arrived on a tide of transformative optimism. Replacing mental hospitals with community-based services was not just going to improve the life of the 'mental patient', it was going to eliminate this unhappy, stigmatized figure entirely. Assisted by the new drugs, people with mental disorders would shake off their outcast status to become normal, or near-normal, family members, workers and citizens. People in crisis would still receive hospital care (in psychiatric wards inside general hospitals), but only until their condition stabilized: no more long-term sequestration in institutional ghettoes. For the minority of very disabled people in need of long-term twenty-four-hour support, small-scale residences would be provided where they too would have freer and happier lives than in the wretched back wards of the asylums.[7]

The guiding principle behind this vision was that the best way of life – for everyone, sick or well – was one of personal independence. 'Dependency' of the kind fostered by the asylum system was condemned as an unmitigated evil; in its stead would come 'effective, self-reliant, productive behaviour'.[8] To this end, UK policymakers insisted that services for the mentally ill should be minimized 'to prevent undue dependency by meeting only the needs the patient is unable to meet through his own efforts'.[9] Services in Canada and the US were similarly designed to encourage individual self-reliance. Open-ended care, inside or outside an institutional framework, was a no-no.

Today this independence imperative is stronger than ever. To need other people on a day-to-day basis – except in the case of the very young, very old or very disabled – is a mark of emotional debility.[10] (In North America it even has diagnostic labels: 'relationship addiction', 'dependent personality disorder'.) This attitude is shared not just by mental health policymakers and professionals but by many service users themselves. This is easily understood. The old asylums, the service-user activist Peter Campbell points out, fostered a 'double dependency': 'On the one hand the users of existing services [were] bred to accept dependency as a characteristic of relationships. On the other, the caring team based their operations on this inequality.'[11] The psychological damage this could cause was dreadful. I saw something of this. One day I asked a group of my creative-writing students to write down their responses to the sentence 'I want to be treated with respect.' As a literary exercise it didn't work well, but it triggered a lively discussion in which most people spoke passionately about their wish to be treated 'just like anyone else'. But one student, the gentle Andy, demurred. People who are 'really sick' are different from other people, he said. 'Sometimes it's right to treat such people like children, because that's what they're like!' I understood what he meant but Andy himself, who had been in psychiatric institutions since late boyhood, was walking testimony to the infantilizing effects of long-term institutionalization, the self-disrespect and lack of confidence bred by the system. 'At times it is hard work not to believe we are a separate branch of humanity,' Peter Campbell observes.[12] For Campbell himself, it was only with his involvement in the service-user movement that he acquired the sense of 'independence and integrity' so long denied to him by the mental health system.[13] Political campaigning brought a different kind of independence, one rooted in mutual support and solidarity. 'I felt stronger as soon as I joined up with other users,' a woman said to me. 'Freer, you know, to be myself . . . with other people around me.'

The service-users' movement (or the survivors' movement, as it also describes itself) originated in the early 1970s. In 1971 a group of New York patients founded the Mental Patients Liberation Project, followed the next year by the Vancouver-based Mental Patients Association. In Britain the movement took off in 1973, during the Paddington Day Hospital fracas when some of the patients caught up in the conflict set up a Mental Patients Union to take forward their demands.[14] It was the age of radical movements and these groups were very typical of their times, issuing calls for a 'mad revolution' to accompany the upcoming 'class revolution'. But the specific issues on which they campaigned – compulsory detention, the right to refuse treatment, patients' representation on policymaking bodies – were important and urgent, and remained so over the decades as other organizations – Campaign against Psychiatric Oppression, Survivors Speak Out, the Hearing Voices Network – emerged and gradually gained a hearing. Voluntary associations acting on behalf of UK service users (notably Mind) had been important players in the mental health arena for decades, but by the end of the twentieth century they had been joined by a large number of independent service-user groups. Today there are hundreds of such groups across Britain, clustered into the National Survivor User Network, a well-organized, lightly professionalized body that works with Mind and other mental health charities to lobby the government. Similar organizations exist in the United States, Canada and many other western countries, with international coordination through the World Network of Users and Survivors of Psychiatry, MindFreedom International, the Hearing Voices International Network and other associations.

I saw almost nothing of the service-users' movement during my years in the mental health system. The only patients' rights campaigner I met was a former Friern resident, an impressive man who later went on to establish a service-users forum in Islington. Neither the Whittington Day Hospital nor Pine Street had any such groups, although the nurses at the day hospital were keen to see a Patients' Union established there. They proposed this at a staff–patient meeting one day, only to be firmly rebuffed by the patients. 'You're on our side, why do we need a union?' I spoke up for it, but only because I could see that Josie, my key nurse, liked the idea and I wanted to curry favour. The idea of doing anything that the staff *didn't* like was unthinkable to me – and my fearfulness was by no means untypical or unreasonable. Even at the day hospital, and even in my case, mental patients were almost powerless. We could be drugged, transferred between institutions, detained in hospital – all without our consent or even our prior knowledge. The popular present-day slogan – 'No decision about us without us' – was a distant dream for service users back in the 1980s.

I found the Whittington Day Hospital to be a liberal, humane environment, and I've no doubt this was true of many other psychiatric institutions at the time. But the scope for mistreatment was enormous. Institutional vulnerability of the sort imposed on mental patients in the late twentieth century was invariably shadowed by potential injury – and there were many places, like Friern, where injury was inflicted.

So are things better now? Reading the titles of the glossy documents pouring out of mental health bodies, one would certainly think so. *Working for Wellness, Mental Health Choices, Living Independently,*

Making Recovery a Reality . . . the PR language is upbeat, the message relentlessly cheery. 'Mental health care [is being] transformed,' the UK Department of Health's *New Horizons* enthused back in 2009. 'Shifting care from hospital to home . . . has made a huge difference in helping people manage their conditions and move on to recovery. The scale and pace of change has been remarkable.'[15]

Rapid change there has certainly been, especially at the organizational level. This has been particularly marked in Britain, where one government-led restructuring after another has left everyone – managers, staff, patients – disoriented and anxious. Have these changes resulted in better patient care? The situation varies from service to service, place to place, so any broad-brush assessment is tricky. But a closer look at the language of mental health policy is revealing, taking us into a Humpty-Dumpty land of keywords – 'recovery', 'wellness', 'choice' – whose meanings, as in the case of Lewis Carroll's prolix egg, depend on the power of those employing them: a topsy-turvy world where it is easy to lose track of the distinction between sense and nonsense, crazy and sane.

Recovery

'Recovery' is everywhere in mental health services today. It was not a word I heard often in the 1980s and it took me some time to unscramble its current meaning. In general, to 'recover' does not mean to get better (which is widely assumed to be impossible in cases of severe mental illness) but to live with a chronic illness as if you do not have it (this is also known as 'wellness' or 'flourishing'). Today across Britain and North America there are 'recovery strategies' and 'recovery teams' and even 'Recovery Colleges', all aiming at helping people to minimize the impact of their symptoms on their daily lives. In theory, this seems like an excellent idea. In the past, people with severe mental disorders were thought to have little capacity for managing their own lives effectively, nor has the mental health system, with some notable exceptions, been geared to such self-management. So reforms that encourage and support individuals in this respect are obviously welcome. But is this what is happening?

The answer would seem to be no. 'Recovery', I was told again and again, has been 'hijacked' for a policy of service cutbacks. The reasons are partly financial (always a 'Cinderella' service, recent years have seen truly savage reductions in mental health budgets[16]) but also programmatic. Mental health care today is a fast-track system geared to getting people back on to their feet, and back into work, as quickly as possible (one critic describes this as the 'garage repair' model of mental health care[17]). People are hustled through a series of time-limited 'interventions'; getting stuck along the way (in hospital, in rehab, in therapy) is anti-recovery. Also anti-recovery are any services that provide open-ended care: thus, all across Britain, day centres have been closed, rehabilitation programmes run down, outpatient services sharply curtailed. North America shows a similar picture. In 1963, in a speech announcing the transfer of mental health services from hospitals to the community, President Kennedy promised to create 1500 community mental health centres across the United States. In the event, only half this number were established and many of these are now closed or threatened with closure. Canadian 'recovery' services are faring no better. Canada is the only major western country with no national mental health strategy, muddling along instead with a patchwork of fragmented, underfunded, uncoordinated services, all suffused with the rhetoric of personal responsibility.[18] Everywhere ongoing care and support are undermined and derogated.

The recovery agenda originated in the late 1980s in the United States, where it migrated into mental health services from anti-addiction programmes like Alcoholics Anonymous. Service-user activists have played a key role in its development.[19] But now some activists are highly critical. In 2012 I joined a service-user Internet discussion forum on 'Survivor History', which should be – although for reasons of confidentiality it cannot be – compulsory reading for anyone concerned with the state of mental health services today. Its contributors are extremely well informed, and their exchanges about the situation of service users, past and present, make for sobering reading. One email string dealt with the recovery strategy and the correlative attack on 'dependency'. A woman with long experience of the system wrote:

> This notion of dependency as the flipside of recovery is really controlling of service users. [It's] one of the justifications for closing centres where, for all that might be wrong with [these] centres, service users have found a sanctuary during rough times and a place to meet with people of similar experience. Deliberately segregating people is one thing; not allowing people to segregate around shared interests on the basis that this creates 'dependency' is quite another.[20]

'This dependency stuff is totally specious,' Anthony Bateman, a leading psychiatrist and medical psychotherapist, said to me. 'It's just an alibi for cutting people loose, leaving them to struggle on their own.' People need other people. True independence – for everyone, well or ill – is rooted in social connection; without this, it is mere isolation and loneliness. This deep need for connectedness is insufficiently acknowledged throughout the whole of our society, not just in the case of people with mental disorders. But the lack of it hits the mentally ill especially hard since it is so often failures of social connection, particularly in early life, that cause such disorders in the first place. 'Recovery', if it is to happen, must redress this lack. Studies of recovery programme outcomes demonstrate very clearly that it is where people feel most strongly supported that the greatest success is achievable.[21] But today in the UK people are shunted about from team to team, person to person, with little continuity. An individual can easily end up dealing with two dozen people or more. Everyone has a 'care coordinator', but close relationships between service users and care coordinators are not encouraged. 'We want people to learn to manage their own situations, not to always rely on somebody else,' I was told by one mental health manager. 'We want people to learn to make their own choices, instead of looking to someone else for the answers.' This sounded all right until I began to think back to my own attempts to 'manage my situation' when I was very unwell, and some of the deadly choices I made at that time. What I needed then, and what I was fortunate enough to get, was effective care. I met with a man who runs a rehab service in London. He complained to me furiously about some support workers who had been using 'choice' as an excuse not to visit people who failed to turn up for things. 'They tell me it's the person's choice – why should they interfere? Someone might be really frightened or depressed, but they don't bother to sit down with them to try and find out; they just say, "Oh, it's his choice!" and ignore them.' Is choice really more important than care?

Choice

Like independence, choice is a must-have of twenty-first-century welfare policy, reflecting western governments' enthusiastic embrace of a supermarket model of public service provision. In 2006 the UK Department of Health initiated a website, 'Our Choices in Mental Health'. Launching the site, a DoH spokesperson declared that 'patients should know that they now have the powers to choose their own path through services and keep control of their lives. They have the preference to choose how, when, where or what treatments they receive.'[22] The following year the government introduced Community Treatment Orders (CTOs), which compel people to take medication under threat of hospital detention if they refuse. CTOs, it was said, would be imposed infrequently and on a short-term basis. In fact in the three years after they were introduced, 14,295 people were placed under CTOs, of which over 70 per cent were still in force by the end of that time period.[23] Black people figure particularly prominently in these figures, making up some 15 per cent of the CTO population (as compared to 2.9 per cent of the general population). Most of the individuals subject to this high level of coercion are not given adequate information about the terms of their CTOs or their rights with regard to them, despite a legal requirement that this should happen.[24]

In the United States, mental health policy wonks extol choice while lawmakers, horrified by the regular gun massacres carried out by people with presumed mental disorders, pass or strengthen laws allowing the courts to impose psychiatric treatment. The Canadian Mental Health Commission identifies choice as the key to effective mental health care, while provincial governments extend court powers to force involuntary hospitalisation or community-based treatment.

And the ironies don't stop here. Once inside the mental health system, most people have no choice about who looks after them, and a high proportion have no say about the type of medication they are given.[25] Today the majority of UK psychiatric inpatients – up to 100 per cent in some inner-city facilities – are in hospital under legal compulsion; in fact in some parts of Britain it is almost impossible to get into hospital on a voluntary basis. (Again, race is a big factor here, with people from black and minority ethnic backgrounds making up a disproportionate percentage of detained patients.) In other words, at a time when institutional incarceration is meant to be a thing of the past, inpatient treatment for mental illness is once again becoming custodial; and this is to say nothing about the high proportion of men and women in prisons who suffer from mental illness. (In Britain it is estimated that over 70 per cent of the prison population have two or more mental disorders; similar estimates are made for the Canadian prison system.[26] In the United States right now, the three biggest mental health providers are prisons.[27]) Wards are overcrowded – some with occupancy rates of 120 per cent – and jammed up with people who don't want to be there. Staff are overstretched; patients over-looked; the atmosphere is often very fraught. The result is an acute-care system described in a recent major report as 'broken and demoralized', with some wards so chaotic and frightening that people badly in need of care will not enter them voluntarily, which sends compulsion rates up further.[28]

This unhappy situation arises partly from bed shortages, especially in the inner cities, but also from a further policy imperative, one so strong that it outweighs clinical need. This is the requirement to min-imize 'risk' — a word that appears far less often in 'public-facing' documents than feel-good terms like 'recovery' and 'choice', but in reality is the main driver behind day-to-day decision-making.

Risk

Risk, an import from the commercial sector, is everywhere in public services today, but it exerts a particularly strong hold over the mental health system, partly as a response to public anxiety about violent patients (much encouraged by the popular media), but more generally because the system, as presently organized, is inherently risky. This is not because people with mental disorders are no longer sequestered in institutions, but because the system has been stripped of the kind of ongoing care relationships necessary to build understanding and trust, which in turn reduce clinical risk.[29] I heard many stories about this. A long-time London psychiatrist told me about his frustration at losing contact with people he had seen regularly for many years, after his outpatient clinic was closed. Now many of his former patients are being seen only by their GPs.[30] A service user with a record of minor violence had been much helped by a psychiatrist whom he saw regularly for over a decade, until she retired; now he is confronted with a different psychiatrist every time he goes to the mental health centre. 'They have no idea who I am, and they don't seem to care much.' And even when the goodwill is there, the time together is so limited, staff turnover often so rapid, that people hardly have a chance to get acquainted. Darren, another service user, had a 'brilliant' community nurse some years ago: 'he was really supportive, he made a big difference'. Recently Darren has had a string of nurses; the one who visits him now 'doesn't listen to me, he just ticks the boxes'. 'I'm just another form to fill in,' a woman writes on a service-user blog.

I met with a senior social worker who runs a recovery team for a mental health trust. Paul reminisced to me about his close relationships with his clients in the past. 'So if somebody got into difficulties, I could just think about what had worked best for them before, and suggest help along those lines. Now all that is gone – you barely know the people you're dealing with, so you just read their "crisis plans".' The imposition of Community Treatment Orders on a growing scale is the flipside of this depersonalization: small wonder 'risk' is a preoccupation of mental health managers everywhere.

People who pose a risk, to themselves or to others, need genuine care, not box-ticking or coercion. The writer Clare Allan is well known for her bestselling novel, *Poppy Shakespeare* (2006), which is based on her experiences as a patient in the Whittington Day Hospital in the late 1990s. In 2009 Allan published a moving tribute to her social worker, who had just retired.[31] She and the social worker, 'Bernadette', had met weekly for eleven years. In the early years of their relationship Allan was a challenging client. She wouldn't speak to Bernadette; she sent her suicide notes; she repeatedly poured boiling water over her own arms and then flaunted them at Bernadette. She yo-yoed in and out of hospital. But Bernadette stuck with her, and over time her resilience, intelligence and warmth allowed Allan to make good use of her.

> With Bernadette's help, I have built for myself a life that feels worth living . . . I still suffer from periods of desperate depression, but Bernadette has guided me through the darkness so many times that I've

at last grown confident of the way, and am confident too that I'll still hear her footsteps beside me when she's gone.[32]

Allan does not need a Bernadette any more, but there are plenty of people who do. The Bernadettes are being pushed out of the system. Anthony Bateman summarized the situation to me: 'The relational, pastoral component of mental health care has been eliminated. All that is left now is a mechanistic, formulaic, depersonalized substitute for quality care.'

Brain Chemistry

What drives all this? The mentality behind our present-day mental health system is a peculiar blend of quick-fix optimism and curative pessimism. Many mental health professionals wax enthusiastic about symptom-reducing remedies – mostly drugs but also some psychological therapies, especially cognitive behavioural therapy (CBT) – while simultaneously endorsing mainstream psychiatry's view of serious mental ailments as brain disorders with limited recovery prospects. For the majority of psychiatrists, any talk about curing mental illness is utopic and anti-scientific, given the 'neurobiological complexity' of serious psychiatric disorders.

'Brain psychiatry', in varying forms, is as old as the psychiatric profession itself. But today it is facing a growing army of well-informed and eloquent critics. The charge sheet is long, headed by paucity of scientific evidence (decades of intensive research, much of it funded by drugs companies, have thus far produced no persuasive evidence for the neurobiological origin of any mental illness[33]); the proliferation of diagnostic categories of dubious validity (newly 'discovered' disorders listed in the most recent edition of the *American Diagnostic and Statistical Manual of Mental Disorders*, the bible of western psychiatry, include shyness ('social anxiety disorder'), grieving over a bereavement for longer than two months ('manic depressive disorder'), and temper tantrums ('oppositional defiant disorder'); and the overuse of drugs which, it is now increasingly recognized, often do more harm than good. Of these, the drugging situation is the one currently attracting most criticism.

For decades, western psychiatrists have been doling out vast quantities of 'anti-psychotics' (the phrasing is revealing), which not only have debilitating side effects but also often heighten or even produce the very psychopathologies that they are supposed to treat, by disrupting brain functioning. Recent studies show an alarming correlation between high medication levels and poor patient 'outcomes'. Countries where anti-psychotics are used far less than in the UK or North America (which has truly terrifying medication levels, not just among people with psychotic illnesses but throughout the entire population, including vast numbers of drugged-up children) show significantly better outcomes.[34] In the UK people with serious mental disorders on long-term medication are dying on average fifteen years earlier than the rest of the population (in Canada it is twenty years earlier, in the USA twenty-five).[35] Data like these, if they related to any other patient group, would be causing an uproar. But if the people are just 'crazies' . . . ?

In fact there are signs of growing concern. A 2007 report from the British Healthcare Commission revealed over-drugging in mental health trusts across the UK and warned the trusts to get their houses in order.[36] In 2012 the editor of the *British Journal of Psychiatry* called for an 'end to the psychopharmacological revolution', arguing that 'the adverse effects of treatment are, to put it simply, not worth the candle'.[37] American and Canadian psychiatrists have issued no similar calls, but recent, much-publicized attacks on drug overuse and Big Pharma's hold over North American psychiatry have aroused public anxiety.[38] Yet no major mental health body in any of these countries has shown any real will to change the situation. 'Community care is built on medication,' Peter Campbell observes.[39] At stake in drugging

is not just the biomedical model of mental illness but the entire struc-
ture of contemporary psychiatry services.

The people who criticize drug therapy are not, with very few
exceptions, arguing against the use of medication *tout court*, but for
its limited and monitored usage on the basis of continuous consultation with the people actually taking the meds – who have been
complaining about 'zombification' and side effects ever since the drugs
were first introduced back in the 1950s. As David Healy, an Irish
psychiatrist and a leading critic of the psychopharmacological indus-
try, points out, if psychiatrists actually listened to their patients instead
of just foisting drugs on them, 'we would have only a few patients on
[drugs] long term'.[40]

Pushing remedies at people instead of listening to them is a very
old story. From eighteenth-century bloodletting and purging, to
twentieth-century psychosurgery, ECT and neuroleptics, the preferred
mode in mind-doctoring has been to deal with madness from a safe
distance. Listening to people, taking their words seriously, brings them
too close. Thinking about the meaning of someone's symptoms –
instead of just trying to dose the symptoms away – risks feelings of
recognition and empathy. Much of western psychiatry is designed to
avoid such feelings. Drugging people on an industrial scale is the
apotheosis of this avoidance strategy. One of the biggest problems
with meds, as one former mental health manager puts it, is that 'they
blunt the affect of doctors',[41] allowing them to imagine themselves as
dispassionate brain-fixers instead of what they really are: fellow human
beings, with all their own anxieties and vulnerabilities, caught up in
complex, highly charged relationships with their patients. Why practitioners would cling to such a fantasy of professional detachment is
no mystery. Madness touches us all in hidden places; the urge to push
it away can be hard to resist. Now it seems most of the mental health
system is yielding to this urge.

I interviewed a number of mental health managers for this book.
Every one of them – despite 'independence', 'recovery', 'choice' and

all the rest of the Orwellian rhetoric – spoke passionately about the need for emotional rapport between service workers and users, and gave examples of up-close, sympathetic care relationships as instances of best practice. This was startling. But what was missing was any acknowledgement of what is required to foster such relationships – that is, reliable, open-ended care, sometimes in supportive institutional environments. Open-ended care breeds 'dependency'; the institutions that once provided it – hospitals, day centres – were 'ghettoes' where people 'silted up': these are now iron orthodoxies. Time is an obsession: every service or treatment must be time-limited (except drugs) lest people begin to rely on them. Even psychological treatments like CBT are strictly time-limited, so as to encourage 'client self-management' and to mitigate 'dependency issues'. There is a fantasy of adult autonomy at work here which is so extreme that it begins to look like casebook accounts of psychotic aloneness, although in this case the anxieties driving the fantasy belong to the professionals rather than the mad.[42] The NHS now offers computerized CBT, so that a person suffering from anxiety or depression can be sorted out without any human interaction whatsoever – and this despite the fact that a host of studies has shown that it is the quality of the relationship with the therapist that determines the outcome of CBT, as it does all therapeutic encounters.[43]

CBT is the modern face of Anglo-American psychotherapy. In North America it is the first choice of family doctors seeking help for anxious and/or depressed patients. In Britain it is the main NHS therapy on offer. In 2006 the British government introduced IAPT, 'Improving Access to Psychological Therapies', a national talking-therapy programme serviced by a mix of NHS, charitable and private providers. Therapies that are chosen to be part of the IAPT programme need to be 'evidence-based' (road-tested in randomized clinical trials) and to have strong track records in getting people back on their feet (and back into work: a great selling point to successive governments) in double-quick time. CBT claims to do this by replacing negative thoughts and self-defeating behaviours with positive thinking and action. It is goal-oriented and 'manualized' (every treatment follows the same steps) and deemed suitable for people with anxiety disorders, depression and, interestingly, schizophrenia, although the number of people with schizophrenia diagnoses currently receiving it is very small. CBT evangelists like the economist Richard Layard, who played a key role in persuading the government to adopt CBT as a front-line treatment for depression, make fabulous claims for its efficacy. CBT will cure 'at least half' those who have recourse to it, Layard says. Government ministers have estimated that up to a million will benefit. A 2009 study of the CBT success rate in Doncaster over a twelve-month period showed that '8 out of 10 people are cured by CBT': a truly impressive – some might say miraculous – result given that most of the Doncaster patients were treated over the phone (another popular CBT delivery method) and that the longest treatment anyone received was two hours and fifteen minutes in total.[44]

In addition to CBT, IAPT offers other 'well-being' therapies, including family therapy, couples therapy and 'dynamic interpersonal therapy' or DIT. DIT was developed recently by a group of UK psychoanalytic psychotherapists for use in the treatment of depression. It is a brief (usually sixteen-session) manualized psychotherapy which aims to help people by training them to 'mentalize', that is, to become more aware of their states of mind and related behaviours and the impact of these on others. Like CBT, DIT describes itself as a 'here and now' therapy rather than a psychoanalytic therapy of the traditional kind. In fact IAPT does not offer traditional psychoanalytic psychotherapies as these are regarded as insufficiently evidenced.[45] Also, such therapies are not manualized and their open-endedness can make them expensive. These are real concerns for any public health system. But reading Layard and other CBT enthusiasts, it seems likely too that the psychoanalytic ethos does not sit well with them. Psychoanalysts are not cheerless folk (at least not the ones I know), but gung-ho optimism is not their style. Freud's famous description of the psychoanalytic aim – 'to transform hysterical misery into common unhappiness' – still resonates in psy-choanalytic circles; 'well-being' promises are not their currency.

CBT has plenty of critics and I'm not interested in adding to their number. I do not doubt that – done intelligently and sympathetically by well-trained therapists – CBT can help some people feel better, at least for a time (studies of its longer-term effects are inconclusive). This is not to be sniffed at. I am delighted that the NHS has embraced psychological therapies as an alternative to drug therapies, as this has undoubtedly thrown open a door which will not easily be closed again. IAPT is underfunded, with long waiting lists, yet at least now every UK citizen is entitled to publicly funded psychotherapy, which is a huge advance on the recent past.

Yet today, all across Britain, adult psychoanalytic therapeutic facili-ties are threatened with closure. An umbrella organization, the British Psychoanalytic Council, declares 'psychotherapy in danger', as long-established residential units and outpatient clinics, forced to com-pete for contracts in the NHS internal marketplace, are losing out to organizations, private as well as public, that provide IAPT therapies, especially CBT. Does this matter? Not if you believe that psychological misery can be relieved quickly, by treating its 'here and now' manifesta-tions while ignoring its deep roots in people's histories. Layard *does* believe this, contrasting the 'forward-looking' approach of CBT to 'backward-looking psychoanalysis',[46] and this anti-psychoanalytic, anti-historical bias can be found throughout the mental health establish-ment. Why look back when to do so is so difficult, so time-intensive, so demanding of both patient and therapist? The arguments roll on as psychoanalytic therapists struggle to defend their approach. But the rejection of history has some larger implications than just the pushback of psychoanalytic therapy – sorry as I am to witness the latter. Because when history goes, so do the people produced by it, whose stories evaporate into a rootless, unbegotten present. This is what is now hap-pening in the British psychiatric system.

One of the most disheartening things I discovered during my

research for this book is that the taking of patients' histories, so long a key feature of clinical psychiatry, is disappearing. It was Dr D who first told me this: 'What you get mostly now are lists of symptoms and recent treatments – hardly any personal information. I ask about the patient's family, whether they have any children, what jobs have they had. Nobody knows. All they care about are symptoms; nobody tries to get to know the person overall.' History-taking used to be a vital diagnostic tool; now it is giving way to symptom checklists and risk assessments. The case history was once the foundation of clinical training; psychiatric textbooks were full of such histories, including lots of reported speech from patients: 'today all one sees [in these books] are statistics and pseudo-mathematical diagrams . . . The individual [patient] has vanished.'[47]

Without history, people disappear. Quizzing my interviewees about this, the enormity of the loss becomes apparent. Unique identities that have taken shape over a lifetime are dissolving into generalities; individuals are becoming diagnostic categories, 'risk profiles', 'crisis plans'. The personal past often weighs heavily on people with mental disorders. Lightening that burden – by revealing it, understanding it, sharing in it – is a vital part of mental healing. A psychiatric system that denies this to people needs to take a hard look at itself, to ponder its own refusals and deficiencies before setting out to remedy the ills of others.

So, to return to my question: what would happen to me, were my difficulties of three decades ago to occur today? My main treatment of choice was a privately funded psychoanalysis, and this option is still open to anyone who can afford it, although the numbers doing so are apparently dwindling. But if I broke down in the course of my psychoanalysis, as I did then? Would I receive the help I needed?

There is a complicated answer to this question and a very simple one. The complicated answer involves crisis teams, acute wards, recovery strategies, care plans . . . a constellation of services that operate fairly well in some parts of the country and much less well in others; maybe I would be lucky.

The simple answer relates to what I needed most: asylum, a safe place to be, a 'stone mother' to hold me for as long as I required it. This I would not get. Friern, the day hospital, Pine Street: all these, and havens like them, are gone. My hostel, I believe, is still there, along with many more like it, but my chances of being housed in such a place would be slim unless I could pay for it – and the fees would be high.[48] But if I were poor, as most people with serious mental disorders are, or eventually become, then I would have to manage as best I could – and this at a time when governments everywhere are reducing welfare payments to the psychologically disabled.

Would I make it? Possibly, especially if I managed to avoid involuntary detainment and treatment, or long-term drugging. But it would be a tough call. People, with or without mental disorders, depend on other people to lead a liveable life: we don't really need history to tell us this, but history shows what happens when we forget it. The mental health system I entered in the 1980s was deeply flawed, but at least it recognized needs – for ongoing care, for asylum, for someone to rely upon when self-reliance is no option – that the present system pretends do not exist, offering in their stead individualist pieties and self-help prescriptions that are a mockery of people's sufferings. The story of the Asylum Age is not a happy one. But if the death of the asylum means the demise of effective and humane mental health care, then this will be more than a bad ending to the story: it will be a tragedy.

NOTES

Prologue

1. 'Desert Island Discs', BBC Radio 4 (16 July 2010).
2. Hilary Mantel, *Giving Up the Ghost* (London, 2010), p. 4.
3. Marcel Proust, *Time Regained* (London, 2000), p. 227.
4. This is discussed further in the Epilogue.
5. For some recent discussions of the language of madness, see: Mikkel Borch-Jacobsen, *Making Minds and Madness* (Cambridge, 2009), Introduction; Darian Leader, *What is Madness?* (London, 2011).

PART ONE

1. History

1. Roland Barthes, 'The Great Family of Man', in *Mythologies* (London, 1981), p. 102.
2. Vladimir Nabokov, *Speak, Memory* (New York, 1989), p. 23.

5. Dead Babies

1. Gwen Adshead, 'Their Dark Materials: Narrative and Identity in Men Who Have Killed', paper to conference 'Narrative, Self-Understanding and the Regulation of Emotion in Psychiatric Disorder', Institute of Philosophy, University of London (12 October 2012).

6. God

1. Hannah Arendt, *The Origins of Totalitarianism* (London, 1994), p. 477.
2. Donald Winnicott, *Playing and Reality* (Harmondsworth, 1982), p. 73.

7. Sex

1. Robert Louis Stevenson, *Virginibus Puerisque*, quoted in Mantel, *Giving Up the Ghost*, p. 208.

9. Filth

1. Colm Tóibín, *The Empty Family* (London, 2010), p. 7.

12. Bad Dreaming

1. Richard Lucas, 'Thomas Freeman', *Newsletter, Association for Psychoanalytic Psychotherapy in the NHS* (28 December 2002), p. 3.

PART TWO

14. The Asylum

1. Films shot in Friern included Lindsay Anderson's *Britannia Hospital* (1982) and Vadim Jean's *Beyond Bedlam* (1994).
2. It became Friern Mental Hospital in 1937, with 'Mental' dropped in the 1950s.
3. Richard Hunter and Ida Macalpine, *Psychiatry for the Poor: 1851 Colney Hatch Asylum–Friern Hospital 1973* (London, 1974), p. 44.
4. J. H. Stallard, 'Pauper Lunatics and Their Treatment', *Transactions of the National Association for the Promotion of Social Science* (1870), p. 465.
5. Hunter and Macalpine, *Psychiatry for the Poor*, p. 45.
6. The sources on which I drew for this account of Friern's history are listed in the Bibliography.
7. Elaine Murphy, *After the Asylums: Community Care for People with Mental Illness* (London, 1991), p. 36; Andrew Scull, *The Most Solitary of Afflictions: Madness and Society in Britain, 1700–1900* (London, 1993), p. 25; Kathleen Jones, *Asylums and After: A Revised History of the Mental Health Services from the Early Eighteenth Century to the 1990s* (London, 1993), Chapter 1.
8. Michel Foucault, *The History of Madness* (London, 2006). I discuss Foucault's critique of the Asylum Age in 'The Demise of the Asylum', *Transactions of the Royal Historical Society*, 21 (London, 2011), pp. 204–7.
9. Scull, *Most Solitary*, pp. 132–8.
10. Ibid., p. 137.

11. Elaine Showalter, *The Female Malady: Women, Madness and English Culture, 1830–1980* (London, 1985), pp. 25–6.

12. Scull, *Most Solitary*, p. 57.

13. Roy Porter, *Mind-Forg'd Manacles: A History of Madness in England from the Restoration to the Regency* (London, 1987), pp. 122–3.

14. Scull, *Most Solitary*, p. 57.

15. Ibid., pp. 69–71; Dana Rovang, 'When Reason Reigns: Madness, Passion and Sovereignty in Late Eighteenth Century England', *History of Psychiatry*, 17:1 (2006), pp. 23–4.

16. For Pinel see Philippe Pinel, *A Treatise on Insanity* (London, 1806); Jan Goldstein, *Console and Classify: The French Psychiatric Profession in the Nineteenth Century* (Chicago, 2001); Dora B. Weiner, '*La Geste de Pinel*: The History of a Psychiatric Myth', in Mark Micale and Roy Porter (eds), *Discovering the History of Psychiatry* (Oxford, 1994), pp. 232–49.

17. Goldstein, *Console and Classify*, p. 88.

18. Quoted in Anne Digby, *Madness, Morality and Medicine: A Study of the York Retreat, 1796–1914* (Cambridge, 1985), p. 170.

19. Moral treatment was already being practised in some English asylums before the founding of the York Retreat, but the Retreat codified it and publicized it to an unprecedented degree: see Leonard Smith, *Cure, Comfort and Safe Custody: Public Lunatic Asylums in Early Nineteenth Century England* (Leicester, 1999), pp. 189–90. For the Retreat, see Samuel Tuke, *Description of the Retreat: An Institution near York, for Insane Persons of the Society of Friends* (1813; Milton Keynes, 2010); Digby, *Madness*; Charles L. Cherry, *A Quiet Haven: Quakers, Moral Treatment and Asylum Reform* (Rutherford, NJ, 1989).

20. Scull, *Most Solitary*, p. 192.

21. Tuke, *Description of Retreat*, p. 168.

22. Andrew Scull, 'Historical Reflections on Asylums, Communities, and the Mentally Ill', *Mentalities*, 11:2 (1997), p. 3. See also Tuke, *Description of the Retreat*, pp. 133–4.

23. According to Samuel Tuke, one of the most important qualifications for an asylum manager was 'a ready sympathy with man, and a habit of conscientious control of the selfish feelings and passions' (quoted in Digby, *Madness*, p. 27).

24. Charles Augustus Tulk, *The Sixty-eighth Report of the Visiting Justices appointed to superintend the management of The County Lunatic Asylum at Hanwell (26 October 1843).*

25. Some early champions of the asylum system claimed that, for many lunatics, removal from family life was a prerequisite to recovery (Scull, 'Historical Reflections', p. 3).

26. Connolly, quoted in Scull, 'Historical Reflections', p. 4.

27. Ibid., pp. 275–6.

28. John Walton, 'Pauper Lunatics in Victorian England', in Andrew Scull (ed.), *Madhouses, Mad-doctors and Madmen: The Social History of Psychiatry in the Victorian Era* (Philadelphia, 1981), p. 168.

29. Harriet Martineau, *A History of the Peace* (London, 1858), Book 3, p. 304; *London Illustrated News* (January 1848).

30. Andrew Scull, *Insanity of Place/Place of Insanity* (Abingdon, 2006), p. 82.

31. Walton, 'Pauper Lunatics', p. 91.

32. Murphy, *After the Asylums*, p. 37.

33. Peter McCandless, 'Curative Asylum, Custodial Hospital' in Roy Porter and David Wright, *The Confinement of the Insane: International Perspectives, 1800–1965* (Cambridge, 2003), pp. 173–192.

34. Scull, *Insanity of Place*, p. 82.

35. Peter Barham, *Closing the Asylum: The Mental Patient in Modern Society* (Harmondsworth, 1997), pp. 75–7; Murphy, *After the Asylums*, pp. 36–41.

36. Hunter and Macalpine, *Psychiatry for the Poor*, p. 24.

37. John Conolly, MD, *A Letter to Benjamin Rotch Esq . . . on the Plan and Government of the Additional Lunatic Asylum for the County of Middlesex, About to be erected at Colney Hatch* (London, 1847), pp. 12, 14, 21.

38. Barham, *Closing the Asylum*, pp. 1–2.

39. Murphy, *After the Asylums*, p. 40.

40. Hunter and Macalpine, *Psychiatry for the Poor*, p. 86; Barham, *Closing the Asylum*, p. 2.

41. Scull, *Most Solitary*, p. 277.

42. Carole Reeves, 'Everyday Life for Jewish Patients in Colney Hatch', *Shemot*, 19:3 (2011).

43. Elizabeth Bott, 'Hospital and Society', *British Journal of Medical Psychology*, 49 (1976), pp. 97, 102, 103, 106; Elaine C. Stewart, 'Community Care in the London Borough of Islington for former short-stay and long-stay patients following the decision to close Friern Hospital', PhD Thesis, University of London, 1999, p. 27.

44. Peter Barham, 'From the Asylum to the Community: The Mental Patient in Postwar Britain', in Marijke Gijswijt-Hofstra and Roy Porter, *Cultures of*

Psychiatry and Mental Health Care in Postwar Britain and the Netherlands (Amsterdam, 1998), pp. 224–6.

45. David Clark, *Social Therapy* (London, 1974); David Clark, *The Story of a Mental Hospital: Fulbourn 1858–1983* (London, 1996).

46. Bott, 'Hospital and Society', pp. 104–5.

47. The full text of Powell's speech is available on www.studymore.org.uk/ xpowell.htm. Powell's decision to close the asylums was strongly influenced by a 1961 study forecasting a steady fall in the need for psychiatric beds (Trevor Turner, 'The History of Deinstitutionalization and Reinstitutionalization', *Psychiatry*, 3:9 (2004), p. 1). In fact, the proportion of the population admitted to inpatient psychiatric institutions grew steadily between 1945 and 1990; what declined was the average length of these admissions, which decreased rapidly from the mid-1950s (James Raftery, 'Decline of Asylum or Poverty of Concept?', in Dylan Tomlinson and John Carrier (eds), *Asylum in the Community* (London, 1996), pp. 18–30; Barham, *Closing the Asylum*, pp. 11–12, 158). The admission rate to Friern Hospital increased by 40 per cent between 1956 and 1961, but with an accelerating shift towards short-stay patients (Letter from Friern Hospital Group Medical Advisory Committee, to Group Secretary, 4 July 1962: LMA: H/12/CH/A/30/1).

48. K. Jones, *Asylums and After*, p. 153.

49. The label 'anti-psychiatrist' was imposed on these campaigners and not used by them, but I employ it here as this is how they are generally described.

50. Peter Barham's *Closing the Asylum* gives an excellent account of the deinstitutionalization of mental health care in the UK. For the UK process in international perspective, see *International Journal of Mental Health*, 11:4 (1982/3) (entire issue); Walid Fakhoury and Stefan Priebe, 'Deinstitutionalization and Reinstitutionalization: Major Changes in the Provision of Mental Health Care', *Psychiatry*, 6:8 (2007), pp. 313–16. For the service-user movement, see Epilogue, n. 14. For R. D. Laing and British 'anti-psychiatry', see Nick Crossley, 'R. D. Laing and the British Anti-Psychiatry Movement', *Social Science and Medicine*, 47:7 (1998), pp. 877–89.

51. Douglas Bennett, 'The Drive towards the Community', in G. Berrios and H. Freeman (eds), *150 Years of British Psychiatry, 1841–1991* (London, 1991), p. 327.

52. Admission rates (including repeated admissions: the notorious 'revolving-door' syndrome) were higher than ever by the end of the 1960s, but these admissions were increasingly short term (see n. 46 above for references).

53. William Sargent, 'Bringing Treatment of Mental Illness into the Twentieth Century', *The Times* (30 July 1974).

54. Barbara Robb, *Sans Everything: A Case to Answer* (London, 1967); Stewart, 'Community Care', pp. 279–80. The uncatalogued files from Friern deposited in the Royal Free Hospital Archives Centre at the time of the hospital's closure include files pertaining to the Robb scandal.

55. Robb, *Sans Everything*. An independent committee of inquiry was established to investigate Robb's allegations of abuse and neglect; for its findings (which many commentators at the time decried as a 'whitewash'), see http://www.sochealth.co.uk/?s=friern. In 1965 the BBC screened a *Newsnight* report about conditions in seven asylums, including Friern which was said to be in very bad repair, with four consultants looking after 500-plus patients each. 'In one ward 47 patients shared 2 baths and 7 sinks. They were bathed without privacy in batches of 12 a night' (*Daily Mail*, 10 September 1965).

56. Peter Sedgwick, *PsychoPolitics* (London, 1982), pp. 192–3. Sedgwick's assessment of the situation was endorsed by a Parliamentary Select Committee, which reported in 1985 that 'the pace of removal of hospital facilities for mental illness has far outrun the provision of services in the community to replace them' (Barham, *Closing the Asylum*, p. xii). 'Any fool can close a long-stay hospital,' the Committee went on to comment, but 'it takes more time and trouble to do it properly and compassionately.'

57. Murphy, *After the Asylums*, pp. 60–85.

58. Stewart, 'Community Care', p. 291.

59. LMA: H/12/CH/A/30/6.

60. Writing about the situation in Friern in 1985, the Chair of its Medical Committee, Rosalind Furlong, described plummeting staff morale: 'When such staff are already working under pressure in adverse conditions, this can have a profound effect' (Rosalind C. Furlong, 'Closure of Large Mental Hospitals – Practicable or Desirable?', *Bulletin of the Royal College of Psychiatrists*, 9 (1985), pp. 130–34).

61. 'In Conversation with Doris Hollander', *Psychiatric Bulletin*, 28 (2004), p. 18; Stewart, 'Community Care', p. 294.

62. House of Commons Debates (6 July 1990) (www.theyworkforyou.com/debates).

63. No one involved in the hospital's closure seems to have made any systematic effort to ascertain the patients' views. Those patients who met with management wanted to see the hospital closed, but they were very anxious about what would become of ex-inmates, and some, like the service-user activist and aca-

demic Diana Rose, became actively involved in supporting patients through the transition into community care (Diana Rose, personal communication).

64. Dylan Tomlinson, *Utopia, Community Care and the Retreat from the Asylums* (Milton Keynes, 1991), p. 135.

65. Tomlinson, *Utopia*; Julian Leff (ed.), *Care in the Community: Illusion or Reality?* (Chichester, 1997); Julian Leff, 'Why is Care in the Community Perceived as a Failure?', *British Journal of Psychiatry*, 179 (2001), pp. 381–3; Tomlinson and Carrier (eds), *Asylum*; Christine McCourt Perring, *The Experience of Psychiatric Hospital Closure* (Avebury, Aldershot, 1993); Barham, *Closing the Asylum*, pp. 21–4.

66. 'Friern Hospital Decommissioning Report', Royal Free Hospital Archives Centre (uncatalogued papers relating to Friern Hospital's closure).

67. Julian Leff, 'The TAPS Project: A Report on 13 Years of Research, 1985–1998', *Psychiatric Bulletin*, 24 (2000), p 165.

68. G. Thornicroft, O. Margolius and D. Jones, 'The TAPS Project 6: New Long-Stay Psychiatric Patients and Social Deprivation', *British Journal of Psychiatry*, 161 (1992), pp. 621–4. Rosalind Furlong, 'Haven within or without the Hospital Gate: A Reappraisal of Asylum Provision in Theory and Practice', in Tomlinson and Carrier (eds), *Asylum*, pp. 158–62.

69. Stewart, 'Community Care', pp. 294–7.

16. In the Bin

1. 'Mental Health Testimony Archive', British Library Sound Archive: John Hart interview; Judith Holt interview.

2. Ibid., John Hart interview.

3. Michael Conran, 'The Patient in Hospital', *Psychoanalytic Psychotherapy*, 1:1 (1985), pp. 31–43.

17. Psychoanalysis and Psychiatry

1. While there is no full-length history of this, there are some publications that I found very useful; these are listed in the notes below and in the Bibliography.

2. Quoted in Malcolm Pines, 'The Development of the Psychodynamic Movement', in Berrios and Freeman, *150 Years of British Psychiatry*, p. 209.

3. Michael J. Clark, 'The Rejection of Psychological Approaches to Mental Disorders in Nineteenth Century Psychiatry', in Scull (ed.), *Madhouses*, pp. 299–300.

4. Graham Richards, 'Britain on the Couch: The Popularization of Psychoanalysis in Britain 1918–1940', *Science in Context*, 13:2 (2000), pp. 183–230.

5. Elizabeth Rous and Andrew Clark, 'Child Psychoanalytic Psychotherapy in the UK National Health Service: A Historical Analysis', *History of Psychiatry*, 20:4 (2009), p. 447.

6. Jeremy Holmes and Richard Lindley, *The Values of Psychotherapy* (Oxford, 1991), pp. 74–7. This contrasted sharply with the situation in the United States where by the 1950s psychoanalysis was the dominant psychiatric treatment, due at least in part to the American movement's policy of permitting only physicians to train as psychoanalysts.

7. D. W. Winnicott, *The Child, the Family and the Outside World* (Harmondsworth, 1981), p. 88.

8. For these precedents, see Rhodri Hayward, *Self Cures: The Transformation of the Psyche in British Primary Care* (forthcoming, 2014), Chapters 1 and 2.

9. Hayward's forthcoming book (see n. 8 above) discusses this 'emotional quarantine' between doctor and patient in relation to older ideas about the curative efficacy of the doctor's moral influence.

10. Denis V. Martin, *Adventure in Psychiatry* (Oxford, 1974), pp. 113–14.

11. Edgar Jones, 'War and the Practice of Psychotherapy: The UK Experience, 1939–1960', *Medical History*, 48:4 (2004), pp. 493–510.

12. K. Jones, *Asylums and After*, pp. 151–2.

13. World Health Organization, *The Community Mental Hospital*, Technical Report Series No. 73 (Geneva, 1953); T. P. Rees, 'Back to Moral Treatment and Community Care', *British Journal of Psychiatry*, 103:431 (1957), pp. 303–13.

14. Martin, *Adventure in Psychiatry*, p. 57.

15. D. Clark, *Story of a Mental Hospital*, p. 180.

16. *Guardian* (August 1984); quoted in D. Pilgrim and A. Rogers, *A Sociology of Mental Health and Illness* (Buckingham, 2001), p. 151.

17. The surprising survivor of the therapeutic-community model is in prison psychotherapy services, where its use is fairly widespread.

18. The Survivor History Group is a small collection of service users and former service users who are compiling a history of the UK service-user movement. For their work, see www.studymore.org.uk. This website, maintained by Andrew Roberts, secretary of the SHG, is a wonderful resource on mental health history.

19. Dorothy Gould, *Service Users' Experience of Recovery under the 2008 Care Programme Approach*, National User Survivor Network and Mental Health Foundation (2012).

20. D. Clark, *Story of a Mental Hospital*, p. 225.

21. Martin, *Adventure in Psychiatry*, p. 137.

22. Anon., 'Patients Speak', in E. Shoenberg (ed.), *A Hospital Looks at Itself: Essays from Claybury* (London, 1972), p. 243.

23. 'Mental Health Testimony Archive', British Library Sound Archive.

24. Ibid., Premila Trivedi interview.

18. Friendship

1. D. Clark, *Story of a Mental Hospital*, p. 163.

2. Interview with Peter Campbell (October 2012). Campbell also refers to this episode in 'Peter Campbell's Story', in Anny Brackx and Catherine Grimshaw (eds), *Mental Health Care in Crisis* (London, 1989), pp. 14–17.

3. Diana Gittins, *Madness in Its Place: Narratives of Severalls Hospital, 1913–1997* (London, 1998), p. 153.

4. Dr Felicity Callard, personal communication.

5. 'Mental Health Testimony Archive', Peter Campbell interview.

19. Mad Women

1. D. Clark, *Story of a Mental Hospital*, p. 34.

2. *The Times* (13 September 1977); Hansard (29 November 1977); *Sun* (12 January 1978); *Barnet Press* (12 May 1978); Tomlinson, *Utopia*, pp. 71–2.

3. Friern was also used as a source of cheap labour; in the early 1980s entire wards worked for a pittance at Standard Telephones (Dr D interview).

4. David Berguer, *The Friern Hospital Story* (London, 2012), p. 55.

5. Showalter, *Female Malady*, p. 81.

6. Ibid.

7. Ibid., pp. 55–6.

8. Lisa Dixon, *Families and Mental Health Treatment* (Washington, DC, 1998), p. 54.
9. R. M. Galatzer-Levy, Louis Kras and J. Galatzer-Levy, *The Scientific Basis of Child Custody Decisions* (London, 2010), p. 134.

20. Day Patient

1. R. D. Laing, *The Voice of Experience* (Harmondsworth, 1982), p. 36.
2. The day hospital was later named after Belle Ridley.
3. Bobby Baker, *Diary Drawings: Mental Illness and Me* (London 2010).

21. The Hostel

1. Medium- and high-secure units are also increasingly provided by private companies: a development which, as two expert commentators noted back in 2003, 'might lead one to the conclusion that "private madhouses" are back, no matter the official names' (Stefan Priebe and T. Turner, 'Reinstitutionalization in Mental Health Care', *British Medical Journal*, 326 (2003), pp. 175–6).

PART THREE

22. Change

1. Adam Phillips, *Missing Out* (London, 2012), pp. 73–4.

23. Separation

1. D. W. Winnicott, 'The Use of an Object and Relating through Identifications', in *Playing and Reality* (Harmondsworth, 1982), pp. 101–11.
2. Ibid., p. 108.

25. Stories

1. J. M. Coetzee, *In the Heart of the Country* (London, 2004), p. 151.

Epilogue: After the Asylums

1. This account of Friern's closure is based on material held in the Royal Free Hospital Archives Centre, and on published sources cited in these notes and listed in the Bibliography.

2. Deborah E. B. Weiner, 'The Erasure of History: From Victorian Asylum to "Princess Park Manor" ', in D. Arnold and A. Ballantyne (eds), *Architecture as Experience: Radical Changes in Spatial Practice* (Routledge, 2003), p. 198.

3. Letter from Colin Rickard, Director of Projects, Hampstead NHS Trust, to M. Langworthy, Lynx Security (18 February 1984), Royal Free Hospital Archives Centre.

4. Weiner, 'Erasure of History', p. 205.

5. *Wall Street Journal* (5 October 2012); *Independent* (9 March 2012).

6. An earlier version of this account of the filming of Princess Park Manor appeared in the *London Review of Books* (8 May 2003) (under the pseudonym 'Eve Blake').

7. John Welshman, 'Rhetoric and Reality: Community Care in England and Wales, 1948–74', in P. Bartlett and D. Wright (eds), *Beyond the Walls of the Asylum* (London, 1999), pp. 205–26.

8. Brackx and Grimshaw (eds), *Mental Health Care*, p. 3.

9. D. Bennett and I. Morris, 'Deinstitutionalization in the United Kingdom', *International Journal of Mental Health*, 11:4 (1983), p. 14.

10. For examples of this policy language see Department of Health, *Independence, Well-Being and Choice* (2005); Department of Health, *From Segregation to Inclusion: Commissioning Guidance on Day Services for People with Mental Health Problems* (2006).

11. Peter Campbell, 'Peter Campbell's Story', in Brackx and Grimshaw (eds), *Mental Health Care*, p. 13.

12. Ibid., p. 15.

13. Ibid., p. 19.

14. Helen Spandler, *Asylum to Action: Paddington Day Hospital, Therapeutic Communities and Beyond* (London, 2006). For the UK service-user movement, see www.studymore.org.uk; Peter Campbell, 'From Little Acorns: The Mental Health Service User Movement', in A. Bell and P. Lindley (eds), *Beyond the Water Towers: The Unfinished Revolution in Mental Health Services, 1985–2005*, Sainsbury Centre for Mental Health (2005), pp. 73–82; Nick Crossley, 'Fish, Field, Habitus and Madness: The First-Wave Mental Health Users Movement in Great Britain', *British Journal of Sociology*, 50:4 (1999), pp. 647–70.

15. Department of Health Mental Health Division, *New Horizons: A Shared Vision for Mental Health*, HMSO (2009), p. 4.
16. In 2012/13 mental health spending in the UK fell in real terms for the second year in a row, following a 1 per cent cut in 2011/12 (*Guardian*, 15 March 2013).
17. Leader, *What is Madness?*, p. 4.
18. Mental Health Commission of Canada, *Toward Recovery and Well-Being*, 2009. For the inadequacy of these services see (among many similar reports) 'Mental Health Care for the Few', *Maclean's*, 22 March 2011.
19. Judy Turner-Crowson and Jan Wallcraft, 'The Recovery Vision for Mental Health Services and Research: A British Perspective', *Psychiatric Rehabilitation Journal*, 25:3 (2002), pp. 245–54.
20. This comment comes from a member of the Survivor History Group; I am grateful to her for permitting me to quote it.
21. William A. Anthony, 'Recovery from Mental Illness', *Psychosocial Rehabilitation Journal*, 16:4 (1993), pp. 531, 533.
22. This website, www.mhchoice.org.uk, appears to be defunct.
23. Care Quality Commission, *Mental Health Act Annual Report: 2011/12*, pp. 77, 82.
24. Ibid., p. 26. 'This level of coercion,' Richard Bentall writes in the *Guardian*, 'damages relationships between patients and services, often leading to greater reluctance to seek psychiatric help during future crises' (1 February 2013).
25. A 2010 survey by the mental health charity Rethink found that about a third of patients had no involvement in the choice of medication.
26. Mental Health Foundation, 'Mental Health Statistics: Prisons': http://www.mentalhealth.org.uk/help-information/mental-health-statistics/prisons/; Centre for Addiction and Mental Health, 'Mental Health and Criminal Justice Policy Framework' (Toronto, 2013).
27. The three prisons are New York's Riker's Island, Chicago's Cook County Jail and the Los Angeles County Jail (http://www.npr.org/2011/09/04/140167676/nations-jails-struggle-with-mentally-ill-prisoners).
28. Schizophrenia Commission, *The Abandoned Illness: Report from the Schizophrenia Commission* (Rethink Mental Illness, 2012), p. 4.
29. John Wilkinson, 'The Politics of Risk and Trust in Mental Health', *Critical Quarterly*, 46:3 (2004), pp. 95–6.
30. In 2012 31 per cent of people with diagnoses of psychosis were being seen in primary care while a further 47 per cent were seen only twice a year by psychiatrists (Schizophrenia Commission, *Abandoned Illness*, p. 68).

31. Clare Allan, 'My Brilliant Survival Guide' (*Guardian*, 14 January 2009).
32. Ibid.
33. Joan Busfield, *Mental Illness* (Cambridge, 2011), pp. 74–105; Robert Whitaker, *Anatomy of an Epidemic* (New York, 2010), pp. 275, 279, 333.
34. Whitaker, *Anatomy of an Epidemic*, pp. 110–11, 308. Evidence for this disparity comes from WHO cross-national studies undertaken between the late 1960s and the early 1980s. More recent studies have challenged these findings, and these studies have then been challenged in turn; the scientific controversy rumbles on.
35. Chin-Kuo Chang et al., 'Life Expectancy at Birth for People with Serious Mental Illness', *Plos One* (18 May 2011); Joe Parks et al., *Morbidity and Mortality in People with Serious Mental Illness*, National Association for State Mental Health Directors (October 2006).
36. Health Care Commission Report (http://news.bbc.co.uk/1/hi/ health/6256185.stm).
37. Peter Tyrer, 'The End of the Psychopharmacological Revolution', *British Journal of Psychiatry* (24 December 2012), p. 168.
38. Joseph Dumit, *Drugs for Life: How Pharmaceutical Companies Define Our Health* (Durham, NC, 2012); Peter Breggin, *Medication Madness* (New York, 2008); Whitaker, *Anatomy of an Epidemic*.
39. Campbell interview (October 2012).
40. Robert Whitaker interview with David Healy, quoted in Whitaker, *Anatomy of an Epidemic*, p. 336; David Healy, *Pharmageddon* (Berkeley, CA, 2012).
41. Wilkinson, 'The Politics of Risk', p. 93.
42. Melanie Klein, 'On the Sense of Loneliness', in *Envy and Gratitude and Other Works, 1946–1963* (London, 1988), pp. 300–313.
43. M. Lambert and D. E. Barley, 'Research Summary on the Therapeutic Relationship and Psychotherapy Outcome', *Psychotherapy*, 38:4 (2001); Paul Gilbert and Robert L. Leahy (eds), *The Therapeutic Relationship in Cognitive Behavioural Therapies* (Hove, 2007). According to one former manager of a mental health service, staff who continue to emphasize the value of the therapeutic relationship are 'dismissed as self-serving and mystifying' (Wilkinson, 'The Politics of Risk', p. 95).
44. http://www.psychminded.co.uk/news/news2009/nov09/Eight-out-of- 10-people-recover-after-CBT003.htm. Accessed 21 January 2012. Richard Layard, D. Clark, M. Knapp and G. Mayraz, 'Cost-benefit Analysis of Psychological Therapy', *National Institute Economic Review*, 202:1 (2007), pp. 90–98.

45. In fact there now exists a substantial body of evidence for the value of long-term psychoanalytic psychotherapies; recent research is summarized in Busfield, *Mental Illness*, pp. 184–5.
46. Richard Layard, *Happiness: Lessons from a New Science* (London, 2005), pp. 8–9. Layard acknowledges however that some severe disorders require deeper psychological interventions.
47. Leader, *What is Madness?*, p. 4.
48. Also, I would be very unlikely to get the same level of personal support as I received at my hostel. Most (voluntary-sector and commercial) providers of supported accommodation provide only 'tenancy support', which handles only practical matters. The support workers delivering this care tend to be poorly qualified and low paid.

SELECT BIBLIOGRAPHY

This is a Select Bibliography of published sources; additional sources, including archive materials, are cited in the notes.

GENERAL HISTORIES OF PSYCHIATRY AND PSYCHIATRIC CARE

G. Berrios and H. Freeman (eds), 150 *Years of British Psychiatry*, 2 vols (London, 1991, 1996)

Joan Busfield, *Managing Madness: Changing Ideas and Practice* (London, 1986)

William Bynum, Roy Porter and Michael Shepherd (eds), *The Anatomy of Madness: Essays in the History of Psychiatry*, 3 vols (London, 1985, 1989)

Michel Foucault, *The History of Madness* (London, 2006; orig. 1967)

Kathleen Jones, *Asylums and After: A Revised History of the Mental Health Services from the Early Eighteenth Century to the 1990s* (London, 1993)

Mark Micale and Roy Porter (eds), *Discovering the History of Psychiatry* (Oxford, 1994)

Roy Porter, *A Social History of Madness: Stories of the Insane* (London, 1987)

——, *Madness: A Brief History* (Oxford, 2002)

Edward Shorter, *A History of Psychiatry: From the Age of the Asylum to the Age of Prozac* (New York, 1997)

Elaine Showalter, *The Female Malady: Women, Madness and English Culture, 1830–1980* (London, 1985)

FRIERN HOSPITAL

David Berguer, *The Friern Hospital Story* (London, 2012)

Richard Hunter and Ida Macalpine, *Psychiatry for the Poor: 1851 Colney Hatch Asylum—Friern Hospital 1973* (London, 1974)

Julian Leff (ed.), *Care in the Community: Illusion or Reality?* (Chichester, 1997)

Christine McCourt Perring, *The Experience of Psychiatric Hospital Closure* (Avebury, Aldershot, 1993)

Barbara Robb, *Sans Everything: A Case to Answer* (London, 1967)

Dylan Tomlinson, *Utopia, Community Care and the Retreat from the Asylums* (Milton Keynes, 1991)

—— and John Carrier (eds), *Asylum in the Community* (London, 1996)

THE RISE AND FALL OF THE ASYLUM

Charles L. Cherry, *A Quiet Haven: Quakers, Moral Treatment and Asylum Reform* (Rutherford, NJ, 1989)

David Clark, *The Story of a Mental Hospital: Fulbourn 1858–1983* (London, 1996)

Anne Digby, *Madness, Morality and Medicine: A Study of the York Retreat, 1796–1914* (Cambridge, 1985)

Marijke Gijswijt-Hofstra and Roy Porter, *Cultures of Psychiatry and Mental Health Care in Postwar Britain and the Netherlands* (Amsterdam, 1998)

Diana Gittins, *Madness in Its Place: Narratives of Severalls Hospital, 1913–1997* (London, 1998)

Erving Goffman, *Asylums* (New York, 1961)

Denis V. Martin, *Adventure in Psychiatry* (Oxford, 1974)

Roy Porter, *Mind-Forg'd Manacles: A History of Madness in England from the Restoration to the Regency* (London, 1987)

Andrew Scull, *Insanity of Place/Place of Insanity* (Abingdon, 2006)

—— (ed.), *Madhouses, Mad-doctors and Madmen: The Social History of Psychiatry in the Victorian Era* (Philadelphia, 1981)

——, *The Most Solitary of Afflictions: Madness and Society in Britain, 1700–1900* (London, 1993)

Vieda Skultans, *Madness and Morals: Ideas on Insanity in the Nineteenth Century* (London, 1975)

Leonard Smith, *Cure, Comfort and Safe Custody: Public Lunatic Asylums in Early Nineteenth Century England* (Leicester, 1999)

Samuel Tuke, *Description of the Retreat: An Institution near York, for Insane Persons of the Society of Friends* (1813; Milton Keynes, 2010)

Sarah Wise, *Inconvenient People: Lunacy, Liberty and the Mad-Doctors in Victorian England* (London, 2012).

In addition to these, for anyone wanting first-hand accounts of asylum life in twentieth-century Britain I strongly recommend the 'Mental Health Testimony Archive' in the British Library Sound Archive.

PSYCHOANALYSIS AND PSYCHOTHERAPY IN BRITAIN

Sally Alexander, 'Psychoanalysis in Britain in the Early Twentieth Century', *History Workshop Journal*, 45 (Spring 1998)

Michael Clark, 'The Rejection of Psychological Approaches to Mental Disorder in Late Nineteenth-Century British Psychiatry', in Scull (ed.), *Madhouses, Mad-doctors and Madmen* (Philadelphia, 1981)

Robert Hinshelwood, 'Psychoanalysis in Britain: Points of Cultural Access, 1893–1918', *International Journal of Psychoanalysis*, 76 (1995), pp. 135–51

Jeremy Holmes and Richard Lindley, *The Values of Psychotherapy* (Oxford, 1991)

Edgar Jones, 'War and the Practice of Psychotherapy: The UK Experience, 1939–1960', *Medical History*, 48:4 (2004), pp. 493–510

Darian Leader, *What is Madness?* (London, 2011)

Jonathan R. Pedder, 'Psychotherapy in the National Health Service: A Short History', *Free Associations*, 6a (1996), pp. 14–27

Adam Phillips, *Winnicott* (London, 1988)

Eric Rayner, *The Independent Mind in British Psychoanalysis* (London, 1991)

Graham Richards, 'Britain on the Couch: The Popularization of Psychoanalysis in Britain 1918–1940', *Science in Context*, 13:2 (2000), pp. 183–230

Mathew Thomson, *Psychological Subjects: Identity, Culture and Health in Twentieth-Century Britain* (Oxford, 2006)

Donald Winnicott, *Playing and Reality* (Harmondsworth, 1982)

COMMUNITY CARE AND THE POST-ASYLUM MENTAL HEALTH SYSTEM

Peter Barham, *Closing the Asylum: The Mental Patient in Modern Society* (Harmondsworth, 1997)

P. Bartlett and D. Wright (eds), *Beyond the Walls of the Asylum* (London, 1999)

Anny Brackx and Catherine Grimshaw (eds), *Mental Health Care in Crisis* (London, 1989)

Joan Busfield, *Mental Illness* (Cambridge, 2011)

Joseph Dumit, *Drugs for Life: How Pharmaceutical Companies Define Our Health* (Durham, NC, 2012)

Mark Hardcastle, David Kennard, Sheila Grandison and Leonard Fagin (eds),
  Experiences of Mental Health In-Patient Care (London, 2007)
David Healy, *The Creation of Psychopharmacology* (Cambridge, MA, 2004)
Elaine Murphy, *After the Asylums: Community Care for People with Mental Illness*
  (London, 1991)
Craig Newnes et al., *This is Madness* (Ross-on-Wye, 1999) and *This is Madness
  Too* (Ross-on-Wye, 2001)
Peter Sedgwick, *PsychoPolitics* (London, 1982)
Barbara Taylor, 'The Demise of the Asylum', *Transactions of the Royal Historical
  Society*, 21 (London, 2011)
Robert Whitaker, *Anatomy of an Epidemic* (New York, 2010)

THE UK SERVICE-USERS MOVEMENT

The best source for this is the excellent website: www.studymore.org.uk.

Peter Campbell, 'From Little Acorns: The Mental Health Service User Move-
  ment', in A. Bell and P. Lindley (eds), *Beyond the Water Towers: The
  Unfinished Revolution in Mental Health Services, 1985–2005*, Sainsbury Cen-
  tre for Mental Health (2005), pp. 73–82
Nick Crossley, 'Fish, Field, Habitus and Madness: The First-Wave Mental Health
  Users Movement in Great Britain', *British Journal of Sociology*, 50:4 (1999),
  pp. 647–70
Helen Spandler, *Asylum to Action: Paddington Day Hospital, Therapeutic
  Communities and Beyond* (London, 2006)

ACKNOWLEDGEMENTS

Many people have helped me with this book. For taking time to talk to me (in some cases on several occasions), my warmest thanks to: Janet Alldred, Bobby Baker, Peter Barham, Anthony Bateman, Anny Brackx, John Bradley, Brendan Byrne, Lesley Caldwell, Felicity Callard, Peter Campbell, Peter Close, Annette de la Cour, Jackie Drury, Dee Fagin, Leonard Fagin, Stephen Fish, Druid Fleming, Rosalind Furlong, Antony Garelick, Ian Griffiths, Doris Hollander, Jeremy Holmes, David Jones, Julian Leff, Geraldine Mason, Elaine Murphy, Neil Palmer, Adam Phillips, Andrew Roberts, Judith Roberts, Diana Rose, Margaret Rustin, Sylvia Tang, David Taylor, Dylan Tomlinson, Bill Travers, Wendy Wallace, John Wilkinson and Peter Wilson. A special thank-you to Joy Dalton for all her kind assistance, then and now.

Sally Alexander, Catherine Hall, Lynne Segal and Cora Kaplan read an early draft of Part I and made helpful suggestions. Kate Hodgkin provided references and good advice. Keith Lowe advised me about the state of mental health services in Canada. Peter Barham brought his own expertise to the historical sections; he also commented helpfully on the Epilogue, as did Annette de la Cour and Peter Close. Felicity Callard, Adam Phillips, Sarah Knott, Poppy Sebag-Montefiore, Tessa McWatt and Lyndal Roper read the whole book in draft and proposed changes that much improved it (although they are in no way responsible for the final result). I am deeply grateful to all these people for their generous help.

Norma Clarke listened patiently as I trawled through my memories, read each chapter as it emerged, and bolstered my courage when it faltered. She also worked her way carefully through two complete drafts: 'thank you' hardly seems adequate, but I offer it anyway. My agent, David Godwin, has been a great supporter of the book since before I even thought of writing it and I am very grateful to him for this, and for his useful comments on the manuscript. Sarah Coward edited the book with tact and grace, and I thank her warmly for this. Simon Prosser, my editor at Hamish Hamilton, has been a wonderful collaborator; it has been a huge pleasure working with him, and the rest of the Hamish Hamilton team.

The Wellcome Trust funded my time off from university duties to write the book and I thank them for this, especially Nils Fietje and David Clayton. My warm thanks also to my colleagues at the University of East London (where I began the book) and Queen Mary University of London (where I completed it) for their support and encouragement.

I thank once again all the good friends who cared for me during my madness years, only some of whom appear (usually under pseudonyms) in the book.

My biggest debts are to 'V' and Lise Henderson – neither of whom had any involvement in the preparation of *The Last Asylum*. I offer it to them now, in the hope that they will forgive me for the liberties I have taken with our shared experiences.

INDEX

Adshead, Dr Gwen xii
aftercare 192
   see also community care
Aftercare Association for Poor and
   Friendless Female Convalescents
   on Leaving Asylums for the
   Insane 192
Allan, Clare, *Poppy Shakespeare*
   (2006) 261
America, asylums in 117
anorexia 195, 196
anti-psychotic drugs *see* drugs;
   tranquillizers
asylums
   admissions to 114, 115, 274 n47,
      274 n52
   American 117
   decline and closure of xi, 116,
      117, 119–21, 148–9, 250–1
   films shot/set in 104, 155,
      249–50, 271 n1, 280 n6
   French 107, 110
   friendships created in 83–4,
      152–56, 157, 159–60
   gender divisions in 148, 164–5,
      173
   history of (pre-twentieth
      century) 104–5, 107–12, 279 n1

mistreatment of patients in
   117–18, 164–8, 254
NHS, integration with (1948) 115
overcrowding, problems of
   112, 113
sexual abuse, cases of 57, 135,
   148, 164–5
tourists, attracting 105–6
violence 111, 118, 162, 164–7,
   169, 170, 174, 260
voluntary admittance to 115,
   117, 259
women, treatment of 167, 173–4
   see also madhouses; private
      mental healthcare

Baker, Bobby 187–8
Barthes, Roland 13, 17–19
Basaglia, Franco 117
'Bedlam' *see* Bethlem Hospital,
   London
Belmont Hospital, London *see*
   Henderson Hospital, London
Bethlem Hospital, London 107, 109,
   173
Bevan, Aneurin 115
Beyond Bedlam (Vadim Jean, 1994)
   271 n1